MOBILE WEB SERVICES

MOBILE WEB SERVICES

Architecture and Implementation

Frederick Hirsch
Nokia

John Kemp
Nokia

Jani Ilkka

John Wiley & Sons, Ltd

Published by *John Wiley & Sons Ltd*, The Atrium, Southern Gate, Chichester, West Sussex, PO19 8SQ, England
Telephone (+44) 1243 779777

Email (for orders and customer service enquiries): cs-books@wiley.co.uk
Visit our Home Page on www.wiley.com

Other Wiley Editorial Offices

John Wiley & Sons Inc., 111 River Street, Hoboken, NJ 07030, USA

Jossey-Bass, 989 Market Street, San Francisco, CA 94103-1741, USA

Wiley-VCH Verlag GmbH, Boschstr. 12, D-69469 Weinheim, Germany

John Wiley & Sons Australia Ltd, 42 McDougall Street, Milton, Queensland 4064, Australia

John Wiley & Sons (Asia) Pte Ltd, 2 Clementi Loop #02-01, Jin Xing Distripark, Singapore 129809

John Wiley & Sons Canada Ltd, 22 Worcester Road, Etobicoke, Ontario, Canada M9W 1L1

British Library Cataloguing in Publication Data
A catalogue record for this book is available from the *British Library*

ISBN-10: 0-470-01596-9 (P/B)
ISBN-13: 978-0-470-01596-4 (P/B)

Typeset by the authors
Printed and bound in Great Britain by Antony Rowe Ltd, Chippenham, Wiltshire
This book is printed on acid-free paper responsibly manufactured from sustainable forestry in which at least two trees are planted for each one used for paper production.

Executive Project Director: Timo Skyttä
Project Manager: Jyrki Kivimäki
Publishing Consultant: Jani Ilkka
Language Revision: Michael Jääskeläinen
Graphic Design and Layout: Juha Pankasalo
Cover Design: Wiley

CHAPTER 1: Introduction
Author: Frederick Hirsch
Reviewed by Greg Carpenter

CHAPTER 2: Introduction to XML
Author: Jani Ilkka
Edited by Frederick Hirsch

CHAPTER 3: Introduction to Service-Oriented Architectures
Author: Frederick Hirsch, Reviewed by John Kemp and Steve Lewontin
Jani Ilkka (WSDL and UDDI), Edited by Frederick Hirsch

CHAPTER 4: Agreement
Author: Michael Mahan (WS-Addressing), Reviewed by Frederick Hirsch
Frederick Hirsch (Policy)

CHAPTER 5: Identity and Security
Author: Frederick Hirsch, Reviewed by John Kemp and Steve Lewontin

CHAPTER 6: Liberty Alliance Technologies
Author: Jani Ilkka, Reviewed by Jukka Kainulainen and John Kemp

CHAPTER 7: Enabling Mobile Web Services
Author: Nokia,
Tapio Kaukonen (Introduction), Reviewed by Frederick Hirsch,
Matti Mattola (the SOA for S60 platform examples), Juha Kaitaniemi and Erkki Solala (Java example).

CHAPTER 8: Conclusions
Author: Frederick Hirsch

CHAPTER 9: Java Client Development
Author: Nokia, Reviewed by Norbert Leser
HelloWS by Nokia
AddressBook by Nokia

CHAPTER 10: C++ Client Development
Author: Nokia, Reviewed by Norbert Leser
Ari Lassila (PhonebookEx section)
Series 80 HelloWS by John Kemp, Norbert Leser, and Matti Mattola
Series 80 wscexample by Matti Mattola
PhonebookEx by Nokia

APPENDIX A: Web Services Standards Organizations
Author: Frederick Hirsch (OASIS, WS-I, W3C)
Björn Wigforss & Timo Skyttä (Liberty)
Reviewed by Frederick Hirsch, Art Barstow (W3C), and Michael Mahan.

APPENDIX B: SOA for S60 platform Quick Reference
Author: Norbert Leser, Reviewed by John Kemp

APPENDIX C: References

USE CASES:
Author: Björn Wigforss

Frederick Hirsch

Frederick Hirsch, a Senior Architect at Nokia, is responsible for the company's Web Services standardization strategy, and for security in Web service products. He is an active contributor to the Web Services Security and Digital Signature Services committees of OASIS, and has worked as an editor in the OASIS Security Services technical committee (SAML 2.0). Frederick is also an editor of the Security Mechanisms specifications in the Technical Expert Group of the Liberty Alliance, and has contributed security sections to the Mobile Web Services architecture and specification of the Open Mobile Alliance. He has also participated in the XML Signature and XML Encryption working groups of W3C, and has co-edited the W3C XKMS requirements. He is Nokia's primary voting representative at OASIS and WS-I, and a member of the OASIS Board of Directors, where he chairs the Strategy Committee. He has extensive experience in distributed systems and security, having worked at AT&T Bell Laboratories, BBN, Open Software Foundation (OSF; later the Open Group Research Institute), and CertCo, as well as a number of smaller companies. He holds degrees from MIT, Stanford, and Boston University.

John Kemp

John Kemp has spent the past two years intimately involved in the development of the Nokia Web services architecture. During that time, he has edited and contributed to several Web service specifications, including the Liberty Alliance Identity Federation and Identity Web Services frameworks, and the OASIS Security Services SAML 2 specification. John has been involved in developing Internet-scale software systems since 1996, helping to build one of the first Web browser–based software applications, the Employease Network. Prior to joining Nokia, John was an independent software developer whose clients included Deutsche Bank and the Liberty Alliance. He holds a degree in Computer Science and Artificial Intelligence, and is now learning to play the ukulele.

Norbert Leser

Norbert Leser started his professional life in hardware engineering in Germany, where he pioneered a government-funded research project for a mobile Unix-based computer. He then joined Siemens to develop networking and security software. Soon afterwards, in 1988, he volunteered to help with the establishment of OSF in Cambridge, Massachusetts. He was one of the leading architects to conceive and integrate the Distributed Computing Environment (DCE), of which several concepts are reoccurring in Web service technologies. After a long journey at OSF, Norbert began to work with startup companies that address the usability of information security. Most notably, he assumed the role of Chief Architect at Liquid Machines in breaking new ground with a highly intuitive and easy-to-use enterprise rights management product line. Norbert joined Nokia's Strategic Architecture group, where he assumed responsibility for bringing Web service technologies to mobile devices in a way that is useful and non-intimidating to users and developers. He currently works specifically on providing guidance for development tools.

The rest of the authors and editors of this book hold various positions at Nokia. Jani Ilkka is an independent publishing consultant.

Table of Contents

Chapter 1: Introduction 1

1.1 Structure of this Book 3

Chapter 2: Introduction to XML 7

2.1 Markup and Elements 8

2.2 Attributes 9

2.3 XML Declaration 9

2.4 Namespaces 9

2.4.1 Namespace Inheritance and Default Namespaces.... 11

2.4.2 Attributes and Default Namespaces 12

2.5 Well-Formedness and Validity 13

2.5.1 Well-Formed XML Documents 13

2.5.2 Validity 13

2.5.3 Document Type Definition (DTD) 13

2.6 XML Schema 14

2.6.1 Datatypes 14

2.6.2 Schema Namespaces and Importing 17

2.7 Summary 20

Chapter 3: Introduction to Service-Oriented Architectures 21

3.1 Service-Oriented Architecture 21

3.1.1 Discovery 22

3.1.2 Description 22

3.1.3 Messaging 22

3.1.4 SOA with Web Services 23

3.1.5 Core Standards 24

3.1.6 Interoperability 25

3.2 Web Services Messaging 26
3.2.1 Simple Object Access Protocol 27
3.2.1.1 SOAP Headers 28
3.2.1.2 Application Payload 28
3.2.1.3 SOAP Intermediaries 31
3.2.1.4 SOAP with Attachments 32
3.2.1.5 SOAP Transport 33
3.3 Web Service Description 34
3.3.1 WSDL Description 35
3.3.2 Technical Details 36
3.3.3 WSDL Document Structure 38
3.3.3.1 Types – XML Schema definitions for Messages 38
3.3.3.2 Message – One-way Message definition 39
3.3.3.3 PortType – message exchange pattern 40
3.3.3.4 Binding – define transport for a message exchange pattern ... 41
3.3.3.5 Port – endpoint defined for MEP bound to transport .. 43
3.3.3.6 Service – Collection of related MEP endpoints 43
3.4 Web Service Discovery 44
3.4.1 UDDI 45
3.4.1.1 UDDI Registry 46
3.4.1.2 UDDI Registry Categorization 46
3.4.1.3 Data Models 47
3.4.2 UDDI Application Programming Interfaces 48
3.4.2.1 Inquiry API 48
3.4.2.2 Publication API 48
3.4.2.3 Subscription API 48
3.5 Summary 49

Chapter 4: Agreement 51
4.1 Web Services Addressing 51
4.2 Background 52
4.3 Purpose of WS-Addressing 52
4.4 Mobile Application Example 53
4.5 WS-Addressing Core 58
4.5.1 Endpoint References 58

4.5.1.1 Dispatch 59
4.5.1.2 Stateful Services and Messaging 60
4.5.1.3 Endpoint Reference Syntax 61
4.5.2 Message Addressing Properties 64
4.6 Binding Specifications 68
4.7 Conclusions 70
4.8 Policy 71
4.8.1 Policy Types 72
4.8.1.1 Authorization Policy 72
4.8.1.2 Domain-Specific Policies 73
4.8.1.3 Policy Interoperability 74
4.8.1.4 Web Service Policy Mechanisms 75

Chapter 5: Identity and Security 81
5.1 Identity and Web Services 81
5.1.1 Authentication – Establishing Identity 82
5.1.2 Authentication Context 85
5.1.3 Web Service Authentication Tokens 85
5.1.4 Replay Attacks 86
5.1.5 Single Sign-On 87
5.1.6 Single Sign-On Enabled Web Services 88
5.1.7 Authorization and Audit 89
5.1.8 Permission-Based Attribute Identity Services 89
5.1.9 Services Enhanced with Identity Information 90
5.1.10 Identity Service Description 91
5.1.11 Identity-Based Discovery Services 91
5.1.12 PAOS 92
5.2 Security and Web Services 94
5.2.1 Security Services and Protocol Layer Interactions 95
5.2.2 Message Integrity 97
5.2.3 Message Confidentiality 100
5.2.4 Mitigating Denial-of-Service Attacks 101
5.2.5 Nature of Mobile Web Services 101
5.3 Summary 102

Chapter 6: Liberty Alliance Identity Technologies........ 103
6.1 Liberty Identity Federation Framework (ID-FF).... 108
6.1.1 Protocols 109
6.1.1.1 Single Sign-On and Federation................................ 109
6.1.1.2 Single Logout 110
6.1.1.3 Federation Termination Notification......................... 110
6.1.1.4 Name Registration Protocol 110
6.1.1.5 Name Identifier Mapping... 111
6.1.2 Profiles 112
6.1.2.1 Single Sign-On and Federation................................ 112
6.1.2.2 Single Logout 113
6.1.2.3 Identity Federation Termination Notification 113
6.1.2.4 Register Name Identifier Profiles.............................. 113
6.1.2.5 NameIdentifier Mapping... 113
6.1.2.6 NameIdentifier Encryption...................................... 114
6.1.2.7 Identity Provider Introduction 114
6.1.3 Bindings 114
6.2 Liberty Identity Web Service Framework (ID-WSF) 115
6.2.1 Authentication Service and Single Sign-On Service Specification.. 116
6.2.1.1 Authentication Protocol.. 116
6.2.1.2 Authentication Service.. 117
6.2.1.3 Single Sign-On Service.. 117
6.2.2 Discovery Service 117
6.2.3 Interaction Service.. 117
6.2.3.1 UserInteraction Element 118
6.2.3.2 RedirectRequest Protocol.. 118
6.2.3.3 Interaction Service 118
6.2.4 Security Mechanisms 119
6.2.4.1 Confidentiality and Privacy 119
6.2.4.2 Authentication 120
6.2.5 Client Profiles 120
6.2.6 Reverse HTTP Binding for SOAP.......................... 120
6.2.7 SOAP Binding... 121
6.2.8 Data Services Template....................................... 121
6.3 Liberty Identity Service Interface Specifications (ID-SIS)... 122

6.4 Summary 122

Chapter 7: Enabling Mobile Web Services.................... 123

7.1 Mobile Web Services Software Development Environment.. 125

7.1.1 Client-Server Architecture.. 126

7.1.2 Active Objects.. 127

7.1.3 Symbian HTTP Stack 127

7.1.4 XML Processing 127

7.2 SOA for S60 Platform Architecture....................... 128

7.2.1 Frameworks... 129

7.3 A Simple Example... 131

7.3.1 Authentication for Basic Web Services 135

7.4 Identity Web Services Support 138

7.4.1 Connecting to an Identity Web Service 138

7.4.2 SOA for S60 Platform Service Database 144

7.4.3 Service Association 145

7.4.4 Identity Providers... 146

7.4.5 Service Matching 147

7.4.6 Service Policies 148

7.5 Web Service Providers.. 148

7.6 SOA for S60 Platform as a Pluggable Architecture.. 149

7.7 SOA for S60 Platform XML Processing................. 150

7.7.1 Introduction to Processing XML Messages 150

7.7.2 SOA for S60 Platform XML API 152

7.7.2.1 CSenBaseElement.. 153

7.7.2.2 CSenDomFragment 154

7.7.2.3 CSenBaseFragment.. 156

7.7.3 Tips on Choosing a Parsing Strategy 158

7.8 Writing Code for Web Service Consumer Applications 158

7.9 Summary 159

Chapter 8: Summary and Next Steps 161

Chapter 9: Java Client Development............................ 163

9.1 Prerequisites.. 164

9.2 Nokia Web Services Framework for Java Architecture.. 164

9.2.1 General Structure of NWSF API 166

9.2.2 Configuration 167

9.3 XML Processing in NWSF 168

9.3.1 Using Element to Create New XML Messages......... 169

9.3.2 Using Element to Parse Incoming Messages........... 170

9.4 HelloWS Example 172

9.5 Connecting to Web Services................................ 175

9.5.1 Connecting to Framework-Based Services 175

9.5.2 Connecting to Web Services without Using a Framework 176

9.6 Using the Service Database................................. 177

9.7 Identity and Security 178

9.7.1 Registering an Identity Provider 178

9.7.2 Associating an Identity Provider with another Service 178

9.7.3 Using IDP-Based Authentication 179

9.7.4 Authenticating SOAP Messages with WS-Security Username Token 185

9.7.5 Self-Authentication with HTTP Basic Authentication... 186

9.7.6 Policy and Services 189

9.8 HelloWS.java 189

9.8.1 HelloWS.java Test Service Servlets 193

9.9 AddressBook Example 197

9.9.1 Environment Overview 198

9.9.2 Building and Running 199

9.9.3 Design and Implementation................................. 199

9.9.3.1 User Interface 199
9.9.3.2 Application Structure 200
9.9.3.3 User Interface Handling 202
9.9.3.4 Service Handling 207
9.9.4 AddressBook.wsdl 215
9.9.5 AddrBook.xsd 216

Chapter 10: C++ Client Development 219
10.1 Nokia Service Development API 222
10.1.1 General Package Structure 222
10.1.2 IDE Usage 223
10.1.3 NWSF XML Library 223
10.1.4 WSDL and NWSF 224
10.2 Hello Web Service Example 226
10.3 NWSF Service Connection Library 230
10.3.1 MSenServiceConsumer Interface 230
10.3.2 HandleMessage() 231
10.3.3 HandleError() 235
10.3.4 SetStatus() 236
10.4 Sending Messages to a Web Service 237
10.4.1 Connecting to a Web Service by Using a Framework 237
10.4.2 Connecting to a Web Service Without Using a Framework 238
10.5 NWSF Service Manager Library 238
10.6 NWSF Service Description Library 238
10.6.1 Using the NWSF Service Database 239
10.7 NWSF and Identity 239
10.7.1 Registering an Identity Provider with NWSF 240
10.7.2 Associating an Identity Provider with a Service 240
10.7.3 Using IDP-Based Authentication 241
10.7.4 Using WS-Security Username Token 241
10.7.5 Using HTTP Basic Authentication 242

10.8 Policy and Services 243
10.8.1 Registering Service Policy with NWSF 244
10.9 Configuration 244
10.10 NWSF Utility Library.. 245
10.11 Description of wscexample Client Code 246
10.11.1 Registering Credentials.................................... 247
10.11.2 Establishing a Service Connection 247
10.11.3 Handling Callbacks 248
10.11.4 Creating a Query 250
10.12 Test Service Code 251
10.13 PhonebookEx Example Application 252
10.13.1 Building the Application 252
10.13.2 PhonebookEx Web Service Consumer 254

Appendix A: Web Services Standards Organizations ... 257
A.1 W3C 257
A.1.1 Value Proposition and Membership Base 258
A.1.2 Organization and Operating Model 258
A.1.3 Deliverables 259
A.1.4 Achievements.. 260
A.1.5 Influence and Future Directions............................ 262
A.2 OASIS 262
A.2.1 Value Proposition and Membership Base 263
A.2.2 Organization and Operating Model 263
A.2.3 Deliverables 264
A.2.4 Achievements.. 264
A.2.5 Influence and Future Directions............................ 266
A.3 WS-I 267
A.3.1 Value Proposition and Membership Base 267
A.3.2 Organization and Operating Model 267
A.3.3 Deliverables 269
A.3.4 Achievements.. 269
A.3.5 Influence and Future Directions............................ 270

A.4 Liberty Alliance.. 271
A.4.1 Value Proposition and Membership Base.............. 271
A.4.2 Organization and Operating Model...................... 272
A.4.3 Deliverables 275
A.4.4 Achievements... 276
A.4.5 Influence and Future Directions........................... 278
A.5 W3C® Document Notice and License 280

Appendix B: Nokia Web Services Development API – Quick Reference.. 283
B.1 Service Connection Library 284
B.2 Service Manager Library...................................... 285
B.3 Service Description Library.................................. 286
B.5 XML Library 290
B.6 Utilities Library.. 294

Appendix C: References... 297
Chapter 2: Introduction to XML 297
Chapter 3: Introduction to Service-Oriented Architectures 299
Chapter 4: Agreement.. 302
Chapter 5: Identity and Security 304
Chapter 6: Liberty Alliance Identity Technologies...... 305
Chapter 7: Enabling Mobile Web Services 310
Chapter 9: Java Client Development......................... 310
Chapter 10: C++ Client Development...................... 311
Appendix A: Web Services Standards Organizations. 312
Appendix B: Nokia Web Services Development API – Quick Reference................................ 313

Index 315

Chapter 1: Introduction

A major industry trend is evident in the deployment of Web services technology to enhance existing services and to create new and innovative services. Web services are being widely deployed to facilitate interoperability across different hardware and software implementations, machine architectures and *application programming interfaces (APIs)*. Such interoperability offers near-term benefits by enabling quicker and cheaper integration of existing services.

It also promises a future where applications can be created by combining multiple services in a single workflow. This will make it easy to adjust application functionality, because services can be added or removed from the application workflow. In addition, interoperability will allow application designers to replace one service implementation with another for technical or business reasons. This vision of *Service-Oriented Architectures (SOAs)* is rapidly becoming a reality through the standardization and deployment of Web services technology.

The increasing use of mobile terminals and infrastructure is another industry trend, one that is making it possible to communicate and access information from any location at any time. The convergence of mobile and Web service technologies enables new services and business models, and accelerates the development of mobile and fixed Internet technologies. The mobile industry is poised to take advantage of the benefits of interoperability that Web services provide. Interoperable message structures can reduce the time and cost of business integration, creating a tremendous business driver for the adoption of Web service technologies. This book provides an introduction to Web service technologies for those familiar with mobile technologies, and information on mobile applications for those familiar with Web services.

The mobile environment poses unique requirements and challenges. Mobile devices, also referred to as terminals in this book, are generally constrained in terms of processing power, battery life and user interface. Mobile networks have different characteristics depending on their implementation, but disconnected operation of terminals is the norm, asynchronous communication is an accepted practice, and limited capacity (bandwidth) and large delay (latency) are to be expected. Today, Web service applications cannot usually access terminals directly.

Mobile device application developers may need to work with a different operating system, such as Symbian, even though they may also work with Java to accelerate the learning curve. This book describes a mobile Web services platform, which enables application developers to work faster and more effectively with mobile devices.

CASE STUDY: TNT PACKAGE TRACKING

The TNT Package Tracking service is another proof-of-concept project by TNT, SysOpen Digia, and Nokia.
This case is an excellent example how basic Web services can be used as the technology of choice for enabling logistics reporting.
In the example, we assume that a subcontractor of TNT needs to report a package delivery status to TNT's package delivery server. In this example, the user captures an image of the package's barcode using the camera phone (such as the Nokia 9500 Communicator). The Web services client in the device then translates the barcode into XML code by means of Optical Character Recognition (OCR), and embeds the results into a SOAP message. The client also provides menus for reporting the status of the package delivery, the shipper's identity, and so on. Finally, a VPN session is set up and the status report is transmitted to the TNT package delivery server as a Web service message.

This use case offers several benefits:

- Easy integration into back-end systems by using end-to-end Web services
- Decreased cost from being able to use off-the-shelf mobile phones
- Real-time, accurate status reports

For the mobile industry, Web services offer at least three major application areas. First, mobile terminals can be made to act as Web service clients, enabling many business and consumer scenarios. For example, applications that allow access from any location to backend databases enable powerful customer relationship management, inventory management, and remote diagnostic applications. Second, mobile terminals can offer Web services to other service providers. For example, a mobile terminal can offer a service providing information stored in the terminal, such as contact, calendar, or other personal information. Third, service providers can leverage information provided by the mobile infrastructure. For example, a service provider can obtain the geographic location of a mobile terminal from a mobile infrastructure Web service. This can be used to provide customized information, such weather or restaurants in or near the terminal user's current geographic location. Other potential areas are related to billing and user presence in the mobile network.

Creating effective mobile Web services requires an architecture that addresses issues related to identity management, security, the machine-readable description of Web services, and methods for discovering Web service instances. This book outlines these issues and related technical approaches. The presentation is made concrete with the discussion of a Nokia software platform designed to simplify the development of Web service applications for mobile terminals. The platform supports core Web services standards, as well as the Liberty Identity Web Services Framework (ID-WSF), which is designed to enable identity-aware Web services. The book outlines core Web services and identity management concepts, and the implementation of the Nokia platform. The book also provides concrete examples and next steps for developers.

With the wide adoption of Web services, the industry is entering a new phase in which more difficult issues are being addressed to bring the technology and related opportunities to the next level. It is clear that additional standards will be needed to address policy, reliability and other issues.

1.1 Structure of this Book

Chapter 2: Introduction to XML
Chapter 2 introduces the Extensible Markup Language (XML), which provides the technological foundation for Web Service technologies. The chapter aims to provide readers with enough information to be able to understand the XML-based features of Web Service technologies.

Chapter 3: Introduction to Service-Oriented Architectures
Chapter 3 provides an overview of key concepts and technologies necessary for understanding the rest of the book. This includes an overview of the concepts related to the Service-Oriented Architecture (SOA) and Web Services, and details related to SOAP messaging, Web service description using WSDL, and discovery using UDDI or a Liberty ID-WSF discovery service.

Chapter 4: Agreement

Chapter 4 describes how policies can be used to govern message exchanges in Web services, as well as the features required in the exchanges. Examples include the need for message correlation, reliable messaging, and the use of appropriate security mechanisms. In addition to the WS-Policy set of specifications, the chapter covers the WS-Addressing specification, which supports message correlation and defines means to specify messaging endpoints.

Chapter 5: Identity and Security

Chapter 5 introduces the concept of identity-based Web services, and the requirements and core technologies related to this concept. The chapter describes how services can be enhanced with identity-based information, and discusses technological aspects, such as the relationship between Single Sign-On and Web services. Discussed technologies include mechanisms to establish identity through authentication, ways to describe the various aspects of an authenticated identity (for example, the quality and type of identity verification used to associate an authentication mechanism with a real identity, and methods for key protection), and the use of security tokens in Web service messages. The chapter concludes with an overview of security mechanisms needed to support Web services, and particularly identity Web services. Such mechanisms include protocol-layer interaction and security options, as well as security services (including integrity and confidentiality).

Chapter 6: Liberty Alliance Identity Specifications

Chapter 6 continues the identity technology discussion started in Chapter 5, and presents a more detailed look at the Liberty Alliance identity specification portfolio. The chapter begins with a description of the three main specification portfolios, followed by an introduction to the main features of each individual specification. The discussion is kept at a general level – the relevant technical details are available in the actual specifications.

Chapter 7: Enabling Mobile Web Services

Chapter 7 introduces two solutions that enable mobile Web services: the SOA for S60 platform and the Nokia Web Services Framework (NWSF) for Series 80. The chapter introduces the platform architectures and their support for extensibility. The support for plug-in frameworks is also discussed, including a detailed description of support for Liberty ID-WSF. The chapter concludes with a detailed description of the platform API and extensibility to support a variety of Web service architectures.

Chapter 8: Summary and Next Steps
Chapter 8 presents a brief summary of all the topics covered in this book, and also provides information on how to start developing Web services by using the SOA for S60 platform.

Chapter 9: Developing Mobile Web Services with Java
While the previous chapters focused on presenting general information and concepts related to mobile Web services, the last two chapters link the concepts to concrete technical information on application development. Chapter 9 describes how to create mobile Web service clients by using Java. The examples range from basic, unsecure Web service messaging to secure, identity-based client applications.

Chapter 10: Developing Mobile Web Services with C++
Chapter 10 follows the pattern presented in chapter 9. At a conceptual level, the C++ application examples are identical to the Java examples. Chapter 10 provides detailed information about the NWSF APIs used in developing mobile Web service clients with C++.

Appendix A: Web Service Standards Organizations
Appendix A provides an overview of relevant standards organizations. It describes the organizations and their structure, as well as their standardization work.

Appendix B: SOA for S60 platform Quick Reference
Appendix B provides a quick reference for the SOA for S60 platform.

Appendix C: References

For more information on Nokia's approach to Web services, visit <http://www.nokia.com/webservices>.

For more information on the Nokia Web Services architecture, visit the Nokia Architecture Update Web site at <http://www.nokia.com/architecture>. Select Technical Architectures, and then Web Services.

For more information on the SOA for S60 platform, visit the Forum Nokia Web site at <http://www.forum.nokia.com>. Select Web Services from the Technologies drop-down menu (on the left under Resources).

Chapter 2: Introduction to XML

At this point, we take a quick detour to cover XML technologies before addressing different aspects of mobile Web services. We do this because virtually all of the Web service technologies presented in the book are based on XML, and consequently rely on the features and restrictions of the XML specifications (there are some exceptions, of course, one being XPath – a non-XML-technique used in the Java examples to define a search string). Readers already familiar with XML may wish to skip this chapter and move straight to chapter 3.

This chapter was written with one goal in mind: to provide the reader with enough information to understand the Web service technology features presented in the book, and to do so without going into too much detail. However, because of its general and introductory nature, many concepts related to XML are not covered in this chapter. For example, we do not discuss element content types and the handling of whitespaces.

The following discussion focuses on the general aspects of XML. A discussion on XML parsers and the intrinsic XML features of the Nokia Web Services Framework can be found in the end of chapter 7.

The Extensible Markup Language (XML) specification and related publications provide a set of specifications defining rules and syntax for creating markup languages.[1] Markup languages are commonly used to describe the contents and structure of text documents. Practical usage examples include presentation (for example, to define the fonts and styles used to display text) and structuring of data documents.

In this book, we focus on the latter aspect: the use of XML to create structured data documents. XML documents are used to store data on computers and mobile terminals and to facilitate the exchange of data between the two.

At the time of writing, the latest XML version is 1.1. The specification became a W3C Recommendation in 2004.

In this section, we present an example of an order message that contains information about the ordered item and the customer making the order. (This message could be augmented by a third-party credit report on the customer.) The merchant processing the order can use the order, customer, and credit information to determine how to process the order.

2.1 Markup and Elements

The actual data or content of an XML document is surrounded by tags that describe the data (Example 2-1). A combination of a start tag, data, and an end tag is called an *element*.[2] The end tag is identical to the start tag except that it begins with a forward slash (and does not contain attributes). For example, an ordered item could be presented in XML as follows:

```
Start Tag         Data                 End Tag
<Name>Mobile Web Services Book</Name>
```

Example 2-1. XML element depicting the name of an ordered item.

An XML element can be nested within other elements. This allows the creation of hierarchical XML data structures. Such hierarchy is demonstrated in the following example, where an XML document provides information about the main authors of this book. Each XML file can only have one root element. In this case, the root element is `OrderMessage`.

```
<OrderMessage>
 <Order>
      <Name>Nokia Mobile Web Services Book</Name>
 </Order>
 <CreditCheck>
    <Provider>
     <Name>Credit Example</Name>
      </Provider>
 </CreditCheck>
</OrderMessage>
```

Example 2-2. XML document with a three-level hierarchy.

2.2 Attributes

In addition to elements, XML documents can store data in *attributes*.[3] Attributes are simply name-value pairs, where a variable is assigned a value. For example, the status of the credit check could be specified with a status attribute as follows:

```
<CreditCheck status="ok">
<Provider>
   <Name>Credit Example</Name>
   </Provider>
</CreditCheck>
```

Example 2-3. Using a `status` *attribute.*

2.3 XML Declaration

Before exploring the functional aspects of XML technology in more detail, one essential element present in virtually all XML documents needs to be introduced: the *XML declaration*.

An XML declaration is placed on the first line of an XML document, and it is used to announce two things: the XML version and the character encoding used in the document.

Since the XML Recommendation version 1.1, the declaration of the XML document version is a mandatory item. An XML document without the declaration is assumed to be based on XML version 1.0. With respect to character encoding, the specification states that all XML parsers must be able to understand UTF-8 and UTF-16 encoding.

```
<?xml version="1.1" encoding="UTF-16"?>
```

Example 2-4. An XML version 1.1 declaration.

2.4 Namespaces

The freedom to define arbitrary tag names can become a problem when two or more parties use XML to exchange information, because different parties could use the same tag to refer to different things. The solution to this problem is to use a technique called *XML namespaces*. XML namespaces are used to qualify element and attribute names by linking a unique resource identifier to each name. The process involves associating a character string, an *Internationalized Resource Identifier (IRI)*, with the tag.[4] This is done by defining an arbitrary short string (called a *prefix*), mapping it to the IRI, and then associating the prefix with tags that are to be associated with that namespace.

To illustrate the tag collision problem and to provide further rationale for namespaces, the following list presents three examples of namespace usage scenarios:

- **Providing modularity by grouping elements into different categories.** One of the benefits of XML achieved through good document design is the ability to reuse the XML structure. Once reusable parts have been placed in unique namespaces, they can be reused in other documents without concern for element or attribute collisions.
- **Combining XML data from multiple sources.** When communicating with several organizations, and especially when combining data from multiple sources, tag name collisions are virtually unavoidable. Let us consider a practical example of a Web store, which combines XML data from different wholesale sources into one XML data structure specific to the Web store. Possible tag name collisions can be avoided, if all data from one source is placed in a source-specific namespace. One example is the independent use of the Name tag for both the order information and the credit check provider. In this case, the credit check can be added independently to the message; the designer of the order schema does not have to anticipate its use.
- **Enabling intermediaries in a complex messaging environment.** In addition to the rather simple example of preventing collisions between identical tags within a single document, namespaces play a larger role in the world of Web services. Namespaces enable the use of SOAP intermediaries: nodes along a message transmission path that may offer services requiring message examination or modification. These SOAP intermediary nodes may need to add, delete, or modify SOAP headers. Use of XML namespaces allows the use of distinct SOAP headers for different purposes, allowing new SOAP headers to be defined and used without concern for potential tag conflicts with headers possibly added by various SOAP nodes.

In addition to the basic feature of qualifying element and attribute names, the Namespaces in XML 1.1 W3C Recommendation contains numerous advanced features to manipulate the way namespaces affect the target XML data. In order to illustrate the namespace-specific features, we present a simplified XML message (Example 2-5). The earlier example showed an order message containing order and credit check information. In that example, it was impossible to distinguish the name of the item from the name of the credit check provider, except from their location in the message. The solution to this problem is to use separate namespaces for orders, customers, and credit checks.

A namespace is declared with an `xmlns` attribute placed on a start tag associating the namespace prefix with an IRI used to create the unique namespace. The prefix is placed in front of the element and attribute names to be associated with the namespace. For the credit check example above, a namespace could be defined as follows:

Order namespace: http://order.example.com/
Credit check namespace: http://creditcheck.example.com/

```
<ord:OrderMessage xmlns:ord="http://order.example.com/">
 <ord:Order>
      <ord:Name>Nokia Mobile Web Services Book</ord:Name>
 </ord:Order>
 <crd:CreditCheck  xmlns:crd="http://creditcheck.example.com">
    <crd:Provider>
     <crd:Name>Credit Example</crd:Name>
      </crd:Provider>
 </crd:CreditCheck>
</ord:OrderMessage>
```

Example 2-5. Using namespaces.

Application developers can freely name the prefix (although some prefixes are reserved, such as *xml*), and can associate it with the appropriate IRI (which must be unique). The IRI does not have to point to an existing object – the validity of the IRI is never checked. In many cases, the IRI is a URL that can be dereferenced (e.g., the URL of a schema file). Multiple namespace attributes can be defined within one element by using multiple `xmlns` declarations.

Attributes do not automatically belong to namespaces. They can also be attached to namespaces individually with attribute-specific prefixes, but this is not usually necessary since processing associated with the element is defined for the attributes as well.

2.4.1 Namespace Inheritance and Default Namespaces

Although namespaces make the XML document structure more modular and hence easier to grasp, they may complicate the interpretation of the document (for human readers) by making it cluttered. To reduce the number of the namespace prefixes, we turn to an advanced namespace technique called *default namespaces*.

A default namespace covers the element it is defined in and is automatically inherited by all the child elements contained within this element. The child elements do not have to be marked with the namespace prefix. All child elements of an element belong to the same namespace, unless specifically defined otherwise. Default namespaces do not associate attributes with namespaces. Default namespaces are defined using a bare `xmlns` attribute, such as `<xmlns="http://order.example.com/">`.

Due to the shortness of our XML document example, namespace defaulting might not appear to offer significant advantages. However, the reduction in the number of prefixes becomes important in longer documents. An example would be to define a default namespace for the order content of our example:

```
<?xml version="1.1" encoding="UTF-16"?>
<OrderMessage xmlns="http://order.example.com/">
 <Order>
      <Name>Nokia Mobile Web Services Book</Name>
 </Order>
 <crd:CreditCheck  status="ok"   xmlns:crd="http://creditcheck.example.com">
    <crd:Provider>
     <crd:Name>Credit Example</crd:Name>
      </crd:Provider>
 </crd:CreditCheck>
</OrderMessage>
```

Example 2-6. Using default namespaces.

2.4.2 Attributes and Default Namespaces

Since default namespaces do not automatically concern attributes, the uniqueness of an attribute should be defined explicitly. However, in practice this is not usually necessary, because attributes are often defined in conjunction with tag definition and thus avoid collisions. In cases where attributes are to be extended with additional attributes, they should be namespace qualified. Continuing our example:

```
<?xml version="1.1" encoding="UTF-16"?>
<OrderMessage xmlns="http://order.example.com/">
 <Order>
      <Name>Nokia Mobile Web Services Book</Name>
 </Order>
 <crd:CreditCheck  crd:status="ok"
xmlns:crd="http://creditcheck.example.com">
    <crd:Provider>
     <crd:Name>Credit Example</crd:Name>
      </crd:Provider>
 </crd:CreditCheck>
</OrderMessage>
```

Example 2-7. Using attribute namespace qualification.

The Namespaces in XML Recommendation defines two absolute requirements regarding the uniqueness of attributes and namespaces in a tag.[5] First, all attributes in a tag must be unique. Second, the complete character string pertaining to an attribute must be unique. In other words, the combination of the attribute and the namespace IRI must not collide with any other attribute.

2.5 Well-Formedness and Validity

This section describes two key concepts of XML technology: well-formedness and validity. They are related to conformance to the rules of the XML specification or a document specifying the document structure and legal data types.

2.5.1 Well-Formed XML Documents

A well-formed XML document meets the constraints specified for well-formed XML documents in the W3C XML Recommendation.[6] In other words, the document conforms to the constraints of XML syntax. To give one example, an XML document must contain only one root element and all other content in the document must be positioned inside the root element.

Another example is that tags must be properly nested. In other words, the element boundaries of an XML document must not overlap. In practice, this means that a start tag of an element must not appear before an end tag belonging to the previous element (obviously, this applies to elements on the same hierarchical level).

2.5.2 Validity

A valid XML document is related to a document defining its structure and the data types used therein. Furthermore, the document must comply with the rules of the document type declaration. Currently, there are two alternatives for the structure and data type definition language: *Document Type Definition (DTD)* and *XML Schema* (both are covered in more detail below). In order to fulfill the validity requirements, an XML document must therefore adhere to either the DTD or the XML Schema definition.

2.5.3 Document Type Definition (DTD)

DTD is a language used to define the document structure for an instance of an XML document. The DTD rules can be found in the XML specification.[7]

The capabilities of DTD are rather limited, and in practice DTD documents can only describe the structure and syntax of the document. At the moment, there are no means to describe the actual data in detail.

2.6 XML Schema

XML Schema contains all features provided by DTD, as well as numerous improvements.[8] With XML Schema, application developers are able to define the instance structure and syntax of XML documents. In addition to a high-level structural definition, XML Schema can be used to define the type of data that a document may contain, all the way down to the smallest detail.

XML Schema is able to handle the actual data types contained by each element and attribute. As a result, it allows the definition of acceptable data types, as well as their limitations. The XML Schema definition contains several built-in data types, but application developers can also create new data types according to their needs.

Application developers familiar with object-oriented programming feel at home with the extensibility and inheritance features of XML Schema. Schema documents can be based on earlier schemas, they can be imported into other schemas, and the inherited rules can be modified or used as such.

2.6.1 Datatypes

Let us begin our tour in XML Schema technology with *datatypes*. They are used to define the exact nature of the data stored in the elements and attributes of an XML document instance. XML Schema types are divided into two groups: simple and complex. The difference is that whereas complex types can contain element content and attributes, the simple types cannot.

Simple Types

The XML Schema specification provides more than forty built-in simple types. The complete list is available in the XML Schema Part 2 document, which also includes an excellent figure illustrating the family tree of the datatypes.[9] The following list describes some common simple types, grouped according to their properties:

– Character strings

  - The `string` type represents a character string. The special feature of this type is that it maintains all whitespace characters, such as tab and carriage return, whereas the `token` data type can contain strings that do not involve, for example, line feed or tab characters.

- Numeric datatypes
    - The `float` type represents a 32-bit floating-point numeric value (single precision).
    - The `decimal` type corresponds to decimal numbers, as the name implies.
    - The `int` type represents normal integer numbers.
- Others
    - The `boolean` type can contain one of two values: true or false.
    - The `base64Binary` type corresponds to Base64-encoded binary data.

Let us put one of the built-in simple types into use in a practical example. To create an element depicting the name of a city, one could use XML Schema as follows:

```
<xsd:element name="City" type="xsd:string"/>
```

Example 2-8. Using a simple schema type.

Although the built-in data type selection is quite extensive, application developers can freely create new simple and complex types. New simple types can be created by restricting the built-in simple types, by creating a list based on simple types, or by combining simple types. The level of detail is evident in that the exact composition of character strings can be defined on a per-character basis with regular expressions.

Complex Types

Before we can move to creating complex types, we need to discuss a certain basic structure used to control the appearance of elements in the actual XML instance. The structure (i.e., the sequence of elements) of a complex type can be defined by using model groups. The application developer adds the desired elements to a model group, and the characteristics of the model group define which elements will be present in the actual document instance, and the order in which they appear. The following three are examples of model groups:[10]

- The elements of the `all` group are all either included in or excluded from the document. If the elements are included in the document, the objects may appear in any order.

- Of all the elements in a `choice` group, the final XML document may contain only one.
- The `sequence` group indicates which elements an XML document may contain. It also dictates their order in the document.

Complex types are defined with a `complexType` element. The example element shown below contains three major parts: the model group, element definitions, and attribute definitions. The following example creates a new complex type called `book`. The example defines two elements called `name` and `author`, and one attribute called `isbn`.

```
<xsd:complexType name="book">
<xsd:sequence>
<xsd:element name="name" type="xsd:string>
<xsd:element name="author" type="xsd:string>
</xsd:sequence>
<xsd:attribute name="isbn" type="xsd:string>
</xsd:complexType>
```

Example 2-9. Defining a complex type called `book`.

The number of element occurrences can be defined with the following attributes:[11]

- `minOccurs` defines the minimum number of the elements in a document. Zero stands for an optional element.
- `maxOccurs` defines the maximum number of the elements in a document.
- If a document contains a `default` attribute, it defines the default value if the instance document does not specify the attribute value.
- If a document contains a fixed-type element, its value must be the one defined by the `fixed` attribute. If the fixed-type element is not included, the XML parser uses the value defined in the fixed attribute.

The occurrence of attributes is controlled with the following options:

- The `use` attribute has three values:
    - `required` – attribute must be included in document
    - `optional` – attribute may be included in document
    - `prohibited` – attribute must not be included in document

- If a `default` attribute is included, its value is the one defined in the document. If the attribute is missing, the default value is used.
- If a `fixed` attribute appears in a document, its value must be the one defined by the attribute. If the `fixed` attribute is not included, the XML parser uses the value defined by the fixed attribute.

Let us use a schema of an Internet bookstore as an example. The example service feature is used to check the quantity of items in the store. The XML message used to convey the quantity-check request message has one parameter: the title of the book. The response message only contains the number of available items. The following XML Schema file defines two elements used in the XML message exchange (the namespaces are explained below.):

```
<schema targetNamespace="http://ji.example.com/bookstore" xmlns="http://www.
w3.org/2000/10/XMLSchema">

        <element name="qtyCheck">
        <complexType>
        <all>
        <element name="nameOfBook" type="string" minOccurs="0"
maxOccurs="1"/>
        </all>
        </complexType>
</element>

        <element name="qtyReply">
        <complexType>
        <sequence>
        <element name="qtyOfBook" type="int" minOccurs="0" maxOccurs="1"/>
        </sequence>
        </complexType>
        </element>

</schema>
```

Example 2-10. Defining two different XML messages.

The `qtyCheck` query message contains one element with `string` content used to identify the book under scrutiny: `nameOfBook`. The number of occurrences is limited to one.

The `qtyReply` response message has one element with `int` content reporting the number of books available; also its occurrence is limited to one.

2.6.2 Schema Namespaces and Importing

Schema documents can be divided into multiple independent documents that may be combined by using the features that enable incorporating external data into an XML Schema instance. (This inclusion feature is a textbook example about the use of namespaces when combining data from multiple sources.) In this section, we first present the namespace system as it pertains to XML Schema documents.[12] After that, we bring external XML Schema data into an XML Schema document by using two different methods: `import` and `include`. We conclude the section by examining techniques pertaining to qualifying locally defined elements.

The definitions and declarations of XML Schema documents are placed in a certain namespace by using the `targetNamespace` attribute (i.e., the namespace shared by all XML document instances adhering to the definitions of the schema document).[13] This attribute plays a major role in the `import` and `include` features explained below. The namespace design relies on the principles explained in section 2.4, while the actual implementation (e.g., the choice of default namespaces) depends largely on the application developer. The following example shows one way to use namespaces with XML Schema documents.

In this example, we take a book quantity enquiry, and add a new element called `order` to illustrate how to create references within a document. The `order` element could be used to inform the store about the name of a title to be ordered, and the quantity of titles bought. First, we set the default namespace to reflect the XML Schema. This saves us the trouble of prefixing all the elements with `xsd`. We also want to specify that this schema document belongs to the bookstore example namespace, hence the `targetNamespace` points to <"`http://ji.example.com/bookstore`">. To facilitate references to the locally defined elements within the document, we create a new `bst` namespace, with a namespace IRI identical to the `targetNamespace`.

```
<schema targetNamespace="http://ji.example.com/bookstore" xmlns="http://www.
w3.org/2000/10/XMLSchema"
xmlns:bst="http://ji.example.com/bookstore">

        <element name="order" type="bst:OrderType"/>

        <complexType name="OrderType">
        <sequence>
        <element name="nameOfBook" type="string" minOccurs="0"
maxOccurs="1"/>
        <element name="qtyOfBook" type="int" minOccurs="0" maxOccurs="1"/>
        </sequence>
        </complexType>

</schema>
```

Example 2-11. An example Schema document with `targetNamespace`.

The `targetNamespace` affects the `import` and `include` features, as `import` is used to bring in data from another namespace, whereas `include` is used to bring in schema definitions from the same namespace.

The `include` element simply references XML Schema files stored elsewhere by using the `schemaLocation` attribute.[14] All files being combined must belong to the same target namespace. For example, if the actual element declarations and type definitions of the schema depicted above were stored in a separate file, the above example would look like this:

```
<schema targetNamespace="http://ji.example.com/bookstore" xmlns="http://www.
w3.org/2000/10/XMLSchema"
xmlns:bst="http://ji.example.com/bookstore">

        <include schemaLocation="http://ji.example.com/bookstore/order.xsd/>

</schema>
```

Example 2-12. An example schema file demonstrating the `include` feature.

The actual schema file named order.xsd looks like this:

```
<schema targetNamespace="http://ji.example.com/bookstore" xmlns="http://www.
w3.org/2000/10/XMLSchema"
xmlns:bst="http://ji.example.com/bookstore">

        <element name="order" type="bst:OrderType"/>

        <complexType name="OrderType">
        <sequence>
        <element name="nameOfBook" type="string" minOccurs="0"
maxOccurs="1"/>
        <element name="qtyOfBook" type="int" minOccurs="0" maxOccurs="1"/>
        </sequence>
        </complexType>

</schema>
```

Example 2-13. Schema file to be included.

Note, that the `targetNamespace` attributes are identical. Although this example brought in a complete schema definition, it might as well have included only parts of a schema. In addition, it is possible to include multiple documents.

Whereas the `include` feature illustrated above works only if all the documents share a common target namespace, the `import` feature can be used to bring in schema definitions aimed at another target namespace.[15] In the following example, the `order.xsd` file belongs to another target namespace. The importing file receives a few changes. First, we define an `ixs` namespace prefix used to refer to the imported schema entities, and then we import the actual schema document called `schemaorder.xsd`.

```
<schema targetNamespace="http://ji.example.com/bookstore" xmlns="http://www.
w3.org/2000/10/XMLSchema"
xmlns:bst="http://ji.example.com/bookstore"
xmlns:ixs="http://import.example.com/">

        <import namespace="http://import.example.com/" schemaLocation="
http://import.example.com/schemaorder.xsd/>

        <!-Refer to the imported types by using the ixs:-prefix -->
        <element ref="ixs:nameOfBook"/>

</schema>
```

Example 2-14. Using `import` to add data from another schema document.

The actual schema file named `schemaorder.xsd` looks like this:

```
<schema targetNamespace="http://import.example.com" xmlns="http://www.
w3.org/2000/10/XMLSchema"
xmlns:imt="http://import.example.com"
elementFormDefault="qualified">



.
.
.
<element name="nameOfBook" type="string" minOccurs="0" maxOccurs="1"/>
.
.
.

</schema>
```

Example 2-15. An example schema document with the element being imported.

The `elementFormDefault` attribute affects the way the locally declared elements and attributes are actually described in the XML document instances.[16] The possible values are `qualified` and `unqualified`, indicating whether or not the entities are qualified with a namespace prefix. With attributes, the respective attribute is called `attributeFormDefault`.

2.7 Summary

The above brief introduction aimed to provide a general understanding of the possibilities that XML and XML Schema offer. A detailed description would easily fill a book of its own. The idea was to present enough information to allow all readers to understand the XML-based Web service examples used in the book. This pragmatic approach inevitably ends up ignoring some issues, such as the content-specific perspective on element definitions.

Chapter 3: Introduction to Service-Oriented Architectures

This chapter introduces key concepts required for understanding Web services. We discuss the importance of interoperability that is based on open standards. We begin with a brief description of the service-oriented architecture deployed using Web services, continue by presenting Web services messaging – particularly Simple Object Access Protocol (SOAP) messaging – and conclude by discussing Web services description and discovery.

3.1 Service-Oriented Architecture

A *service-oriented architecture (SOA)* describes an environment where software applications expose functionality via a service provider messaging interface. Other software agents can act as service consumers in using the functionality exposed by service providers.

Software applications can act as both *service consumers* and *service providers* in providing functionality to the user. For example, a travel agent's phone application can assist the user in planning a trip by accessing an airline service (to book a flight) and a hotel reservation service (to book a hotel room). Thus, the phone travel software consumed two services in order to provide one service (booking a trip). The service-oriented world is not limited to any particular set of devices. Anything that can run a software application qualifies: mobile phones, personal computers, or even home heating systems can all provide and consume services.

As outlined by the World Wide Web Consortium (W3C), a SOA contains the following three major components:[1]

1. Discovery
2. Description
3. Messaging

3.1.1 Discovery

Before one party (requester) can contact another (provider) to obtain services, the requester must locate an appropriate provider. This process is known as *discovery*. Different mechanisms can be used in the process, such as finding services listed on a Web site, searching a directory, or locating the provider through a framework that supports discovery. Different discovery techniques are covered in more detail in section 3.4.

3.1.2 Description

Once the provider has been located, the requester needs to learn the rules related to contacting the provider (e.g., which protocols to use, how to structure messages, and which security and other policies to observe). This process is known as *description*. The description information can be returned as part of the discovery process. The description typically includes structured information, such as Web Services Description Language (WSDL) documents, as well as other information. Issues related to Web services description are described in more detail below in section 3.3.

3.1.3 Messaging

The third step is to communicate with the provider by sending (and possibly receiving) messages. The communication usually takes place using the SOAP protocol together with underlying protocols such as HTTP. The definition of how an underlying protocol is to be used with SOAP is called a SOAP binding. It is required in order to achieve interoperability between SOAP messages and their underlying protocol. Messaging protocols and technologies are described in more detail in section 3.2. The major components of the architecture are illustrated in Figure 3-1 below.

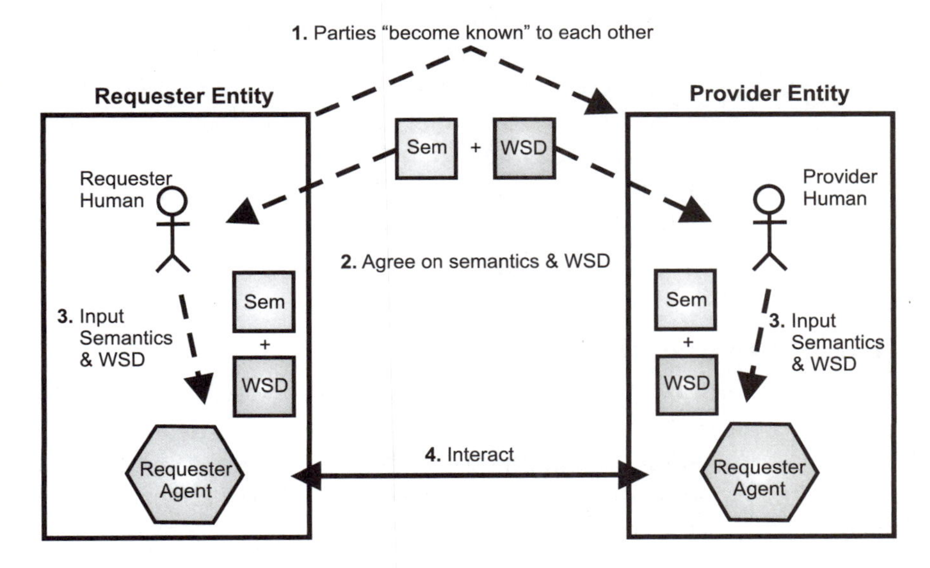

Figure 3-1. Major components of a service-oriented architecture. Source: W3C, Web Services Architecture. See copyright note in Appendix A.

Figure 3-1 shows that in addition to WSDL descriptions, we need semantics. Currently they are described in business agreements, but work on representing semantics in machine-readable form by using Semantic Web technologies is underway.

3.1.4 SOA with Web Services

Web services refer to a specific set of technologies that can be used to implement a SOA. Although SOAs can also be created by using other technologies, Web services are being widely adopted. One reason for the success of Web services is the immediate benefit they bring to systems integration projects. Web services technology can be used to wrap legacy systems and reduce the cost of integration, which produces immediate financial advantages. These advantages include the elimination of duplicated data, avoidance of the errors and costs involved in re-entering data, as well as the competitive advantages of integrating systems, resources, and information. The SOA allows additional functionality to be composed using Web service messages as required. When compared to earlier development efforts, SOA has enabled simple and relatively inexpensive launching of services, and more extensive future functionality.

The description aspect relies on Web Service Description Language (WSDL), which is an XML language used to describe message endpoints and how they should be communicated with. A major advantage of WSDL is its machine-processable format, which enables automation in establishing communication between network endpoints. This can reduce costs by eliminating the need for custom client coding.

Another key technology used by Web services is Universal Description, Discovery and Integration (UDDI), a directory service for locating services by name, type, or other criteria. While a Web service can be implemented without using WSDL or UDDI, these technologies can improve Web service implementations considerably where applicable.

3.1.5 Core Standards

The core standards behind Web services, such as XML, HTTP, SOAP, WSDL, and XML Security, have been standardized by the W3C. Additional standards, such as UDDI, ebXML, SOAP Message Security, and the Security Assertions Markup Language (SAML), have been developed by the Organization for the Advancement of Structured Information Standards (OASIS). SOAP and WSDL have been profiled for interoperability by the Web Services Interoperability Organization (WS-I), resulting in the WS-I Basic Profile. The organization is currently profiling SOAP Message Security. Finally, some of these open standards have been adopted by the Open Mobile Alliance (OMA) in its Mobile Web Services Working Group as the core of specifications aimed to support Web service application designers. The interoperability of these core technologies is based on open standards.

The core standards related to Web services focus on messaging, description, and directories that allow the development of implementations that are independent of computing platform architecture, operating system, computer language, and programming interface. However, this does not prevent the related standardization of programming interfaces for specific platforms from supporting Web services.

One example is the JCP standardization effort of the Java community. The Java language has a large developer base. It has also been widely adopted in both personal computers and mobile devices. Java was designed with device and platform independence in mind, even though certain practical considerations have led to some variation in Java deployments. Specific activity in Java standardization includes proposals to define standard Java interfaces related to Web services, Web service security, and XML security. Examples include the JSR community process J2ME proposals for APIs related to XML Signature and XML Encryption (JSR 105 and 106), J2ME Web Services APIs (JSR 172), an API for Web service message security (JSR 183) and an API for Web service security assertions (JSR 155 for SAML). Of these, JSR 105 involving XML Signature has progressed to a proposed final draft. One of the latest attempts to standardize a Java API for connecting to XML-based services in JCP is called JSR 279. It defines a new high-level API for connection services via frameworks supporting identity based services, discovery, and authentication. The API supports Service Oriented Architectures (SOA) and other similar network service application models. There is also additional activity related to J2SE and J2EE.

3.1.6 Interoperability

The main reason for the success of Web services is that the method of enabling interoperability is not a standard programming interface but a standard format for a messaging protocol (a format for the wire, as it were). This allows the creation of loosely coupled systems regardless of the programming language, hardware platform, operating system, or development environment. The only requirement is that the systems can produce and consume standard protocol messages. Such a loosely coupled system is modular, which makes it easier to modify individual parts of the system independently (for example, to add new services to a message flow, replace service implementations, or even reconfigure entire applications). As result, it is possible to achieve interoperability in a wide range of systems, with minimal impact on system implementations.

One example is the Supply Chain Management Sample Application Architecture published by WS-I. In this example, one or more retail systems access one or more warehouses that replenish their stock from one or more manufacturers. WS-I demonstrates the application's interoperability by having different vendors provide different implementations for the retailer, warehouse, and manufacturer systems, all running as services on the Internet. The application allows different combinations to be selected for demonstration to highlight interoperability. The architecture is illustrated in Figure 3-2 below.

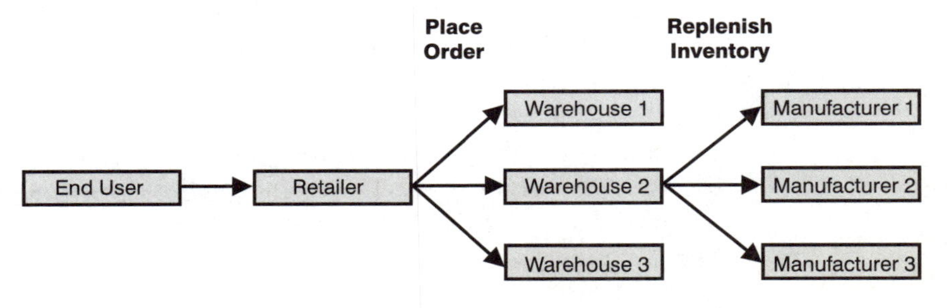

Figure 3-2. WS-I Supply Chain Management Sample Application Architecture.

SOAs allow systems to communicate by using a standard message format and standardized message headers for security and other control functions. They define a messaging structure that supports the composition of common messaging features, such as security and reliable messaging. As a result, development frameworks can support the composition of functionality with the assurance of basic interoperability for the messaging features. It also means that application developers do not have to define how information is to be carried. SOAs also support the future possibility of dynamically creating and managing workflow systems. This will allow service providers to dynamically modify their value chains and service functionality in order to outdo competitors offering similar services.

SOAs enable new business models and offer the potential of reduced costs and increased flexibility. Traditionally computer systems have been developed as stand-alone systems for specific functions, such as payroll or inventory. It has been a major business goal to integrate different systems – to remove the silos. At first, this was done by using custom integration projects and proprietary communication mechanisms or systems designed to centralize the integration. In general, it is very expensive to integrate systems that were not designed to work together, because the integration requires custom design and development, as well as testing and problem resolution. The SOA model solves the integration and interoperability problem by using standard external messaging interfaces that allow systems to interconnect without requiring extensive modification or custom development projects.

3.2 Web Services Messaging

Since service-oriented applications are created from components that send messages to each other, messaging forms a core aspect of Web services. Other aspects can be handled out of band, whereas the messaging component is critical. The adoption of Web services began from simple applications requiring messaging in order to achieve interoperability. When getting started, Web service developers can put off issues related to description, discovery, security, and reliability, and address them when necessary.

Messaging requires a method for addressing messages to the correct endpoints. A URL is part of the addressing information, but often the receiver requires additional information to dispatch and process the message properly. Addressing is discussed in detail in chapter 4. The following sections focus on the SOAP Web services messaging protocol.

3.2.1 Simple Object Access Protocol

Simple Object Access Protocol (SOAP) defines a messaging envelope structure designed to carry application payload in one portion of the envelope (the message body) and control information in another (the message header). The SOAP header can contain any number of SOAP header blocks, which can be used to meet different messaging requirements, such as addressing, security, or reliable messaging. The SOAP body can contain one or more XML element blocks, text, or no content. Binary content can be encoded as text and conveyed in the SOAP body or sent as an attachment, as described below. The general structure of a SOAP envelope is illustrated below in Figure 3-3.

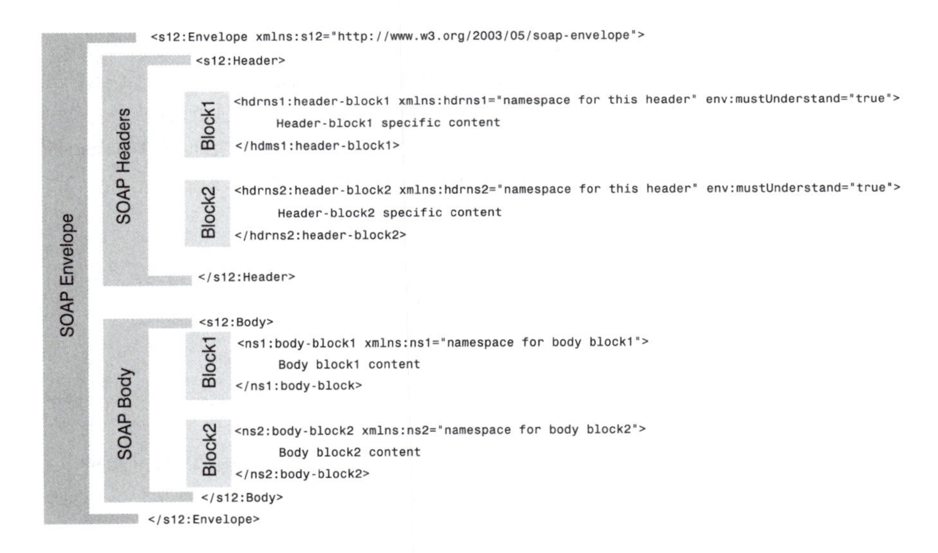

Figure 3-3. SOAP envelope structure.

As an application of XML, the SOAP envelope structure is designed to be extensible. It can be extended with headers defined in XML, and it can convey XML payload. XML namespaces prevent collisions among SOAP header block and body payload elements. SOAP enjoys the benefits of XML, such as the self-describing structure, human readability (useful for debugging) and extensibility. In addition, the experience and tools developed with XML can also be leveraged with SOAP. The self-describing structure enables loosely-coupled modules to interoperate as long as they understand the processing rules associated with specific headers. SOAP avoids the problems of earlier binary protocols such as the *Electronic Data Interchange (EDI)*, in which extensibility required the reuse of binary fields and caused confusion about which agreement was in effect for a given message.

3.2.1.1 SOAP Headers

SOAP header blocks can be added to and removed from the header portion of a SOAP envelope. The blocks are targeted at specific nodes in the SOAP message flow. In SOAP 1.0, these nodes were known as actors, while in SOAP 1.2, they are referred to as roles. SOAP headers can be used for different purposes (such as security, messaging reliability, addressing, and billing), and can be processed by independent SOAP nodes. One of the major advantages of SOAP is that the headers are composable. This allows a variety of application requirements to be met independently with the careful use of SOAP headers. The ability to adjust to changing requirements quickly and easily is a valuable benefit in composing headers when compared to reworking an entire messaging structure.

The SOAP security header is a good example. This particular header is defined in the OASIS Web Services: SOAP Message Security 1.0 (WS-Security) standard. Example 3-1 below describes a simple `UsernameToken` element, which is carried in the security header to convey username and password information.

```
<wsse:Security xmlns:wsse="…">
<wsse:UsernameToken>
<wsse:Username>Tony</wsse:Username>
<wsse:Password>aPassword</wsse:Password>
</wsse:UsernameToken>
</wsse:Security>
```

Example 3-1. Example of a SOAP security header containing a username token.

In this case we assume that the message is transported over a secure transport layer such as SSL/TLS, which provides peer authentication, message integrity, and message confidentiality. This ensures that the application-level authentication data is conveyed securely (when no SOAP intermediaries are present). The OASIS Web Services Security: Security Username Token Profile standard provides a number of tokens that can be used to secure a message and to convey security information end-to-end. Security and identity are vital for Web services and are discussed in more detail in chapter 5.

3.2.1.2 Application Payload

To continue with the same example message, the use of an envelope structure allows an application to send content (payload) within the body of a SOAP message, and to send control information in the header. The header blocks can provide message security and other functions without involving the application. The application is responsible for structuring the payload of a message sent using Web services. Typically the payload is structured in XML because of the advantages the language offers. SOAP supports two operating models: *document messaging* and *Remote Procedure Call (RPC)*.

The document messaging model is being widely adopted because it imposes few constraints on the endpoints, matches many existing business processes and is well suited for a workflow model that may include intermediaries. For example, a purchase order document can be routed among processing nodes by means of a SOAP message. In the RPC SOAP model, parameters associated with function calls (that are made remotely using SOAP) can be formatted automatically. This method is useful in situations where systems are more tightly coupled.

SOAP also defines an encoding method used to resolve certain ambiguities. This feature is most useful in the RPC model. Usually the document model is used with document literal encoding. This means that the content of the body is sent from the initial SOAP sender to the ultimate SOAP receiver, and the sending and receiving applications are responsible for interpreting and processing this content. An application can define the structure of the payload content using XML Schema. This allows a receiver with access to the schema to automatically validate the content structure and element types as part of the parsing process, without requiring special application code to detect basic formatting errors (for example, a zip code type can be enforced).

A SOAP message structure can be defined using XML Schema, and a specific SOAP message instance can be written by using SOAP elements and attributes that comply with that schema.

As an example of SOAP payload we use a SOAP request used to obtain a weather report for a specific city (the service is available at <http://www.webservicex.net/>). The schema for such a request payload is simple, since it consists of two optional strings contained within a specific element. The response in this case is one or more result elements containing string content. The schema is illustrated in Example 3-2 below.

```
<s:schema elementFormDefault="qualifi ed" targetNamespace=http://www.
webserviceX.NET xmlns:s="http://www.w3.org/2001/XMLSchema">
<s:element name="GetWeather">
<s:complexType>
<s:sequence>
<s:element minOccurs="0" maxOccurs="1" name="CityName"
type="s:string" />
<s:element minOccurs="0" maxOccurs="1" name="CountryName"
type="s:string" />
</s:sequence>
</s:complexType>
</s:element>
<s:element name="GetWeatherResponse">
<s:complexType>
<s:sequence>
<s:element minOccurs="0" maxOccurs="1" name="GetWeatherResult"
type="s:string" />
</s:sequence>
```

```
</s:complexType>
</s:element>
</s:schema>
```

Example 3-2. XML schema for weather request and response.

The corresponding SOAP request and response include a payload within a SOAP envelope. In this example, the messages are sent over HTTP, as shown in Example 3-3 below.

```
Request:
POST /globalweather.asmx HTTP/1.0
Host : www.webservicex.net
Content-Type : text/xml; charset=utf-8
Content-Length : 405
SOAPAction : http://www.webserviceX.NET/GetWeather
<?xml version="1.0" encoding="utf-8"?>
<soap:Envelope xmlns:xsi="http://www.w3.org/2001/XMLSchema-instance"
xmlns:xsd="http://www.w3.org/2001/XMLSchema" xmlns:soap="http://schemas.
xmlsoap.org/soap/envelope/">

<soap:Body>
<GetWeather xmlns="http://www.webserviceX.NET">
<CityName>Boston</CityName>
<CountryName>United States</CountryName>
</GetWeather>
</soap:Body>
</soap:Envelope>

Response:
200 OK
Connection: close
Date: Fri, 12 Nov 2004 12:12:35 GMT
Server: Microsoft-IIS/6.0
X-Powered-By: ASP.NET
X-AspNet-Version: 1.1.4322
Cache-Control: private, max-age=0
Content-Type: text/xml; charset=utf-8
Content-Length: 1211
<?xml version="1.0" encoding="utf-8"?><soap:Envelope xmlns:soap="http://
schemas.xmlsoap.org/soap/envelope/" xmlns:xsi="http://www.w3.org/2001/
XMLSchema-instance" xmlns:xsd="http://www.w3.org/2001/XMLSchema"><soap:
Body><GetWeatherResponse xmlns="http://www.webserviceX.NET"><GetWeatherRes
ult>&lt;?xml version="1.0" encoding="utf-16"?&gt;
&lt;CurrentWeather&gt;
&lt;Location&gt;Boston, Logan International Airport, MA, United States
(KBOS) 42-21-38N 071-00-38W 54M&lt;/Location&gt;
&lt;Time&gt;Nov 12, 2004 - 06:54 AM EST / 2004.11.12 1154 UTC&lt;/
Time&gt;
&lt;Wind&gt; from the NE (050 degrees) at 15 MPH (13 KT):0&lt;/Wind&gt;
&lt;Visibility&gt; 10 mile(s):0&lt;/Visibility&gt;
&lt;SkyConditions&gt; overcast&lt;/SkyConditions&gt;
&lt;Temperature&gt; 39.9 F (4.4 C)&lt;/Temperature&gt;
&lt;DewPoint&gt; 25.0 F (-3.9 C)&lt;/DewPoint&gt;
&lt;RelativeHumidity&gt; 54%&lt;/RelativeHumidity&gt;
&lt;Pressure&gt; 30.39 in. Hg (1029 hPa)&lt;/Pressure&gt;
&lt;PressureTendency&gt; 0.02 inches (0.8 hPa) higher than three hours
ago&lt;/PressureTendency&gt;
&lt;Status&gt;Success&lt;/Status&gt;
&lt;/CurrentWeather&gt;</GetWeatherResult></GetWeatherResponse></soap:
Body></soap:Envelope>
```

Example 3-3. SOAP request and response for a weather report Web service.

In our example, the weather report is returned as a single encoded string, which can be decoded to produce the XML used to deliver the weather report. The XML code could have been produced without the declaration, but this method removes any concern regarding namespaces. The decoded weather report is shown in Example 3-4 below.

```
<?xml version="1.0" encoding="utf-16"?>
<CurrentWeather>
<Location>Boston, Logan International Airport, MA, United States (KBOS)
42-21-38N 071-00-38W 54M</Location>

<Time>Nov 12, 2004 - 07:54 PM EST / 2004.11.13 0054 UTC</Time>
<Wind> from the N (010 degrees) at 12 MPH (10 KT):0</Wind>
<Visibility> 1 3/4 mile(s):0</Visibility>
<SkyConditions> overcast</SkyConditions>
<Temperature> 32.0 F (0.0 C)</Temperature>
<Wind>Windchill: 23 F (-5 C):1</Wind>
<DewPoint> 30.0 F (-1.1 C)</DewPoint>
<RelativeHumidity> 92%</RelativeHumidity>
<Pressure> 30.42 in. Hg (1030 hPa)</Pressure>
<Status>Success</Status>
</CurrentWeather>
```

Example 3-4. Decoded weather report.

3.2.1.3 SOAP Intermediaries

The SOAP protocol is different from traditional client-server protocols in that it explicitly supports message flows with intermediaries. This means that a single SOAP message can be processed by SOAP nodes located between the initial SOAP sender and the ultimate SOAP receiver, as well as by the ultimate SOAP receiver.

Intermediary nodes can add or remove SOAP headers and perform additional processing, as defined in the SOAP processing rules. This enables business models impossible in the traditional client-server model, because the messaging model used in SOAP is closely aligned with a workflow business model.

One example is a situation where all messages sent between a corporation and its business partner must be signed using a corporate key and must have the payload encrypted for confidentiality. This can be achieved by using a corporate security gateway, which ensures that all incoming SOAP messages have been correctly signed by the partner. The gateway also signs all messages destined for the partner. In addition, the gateway may encrypt and decrypt message content at the corporate boundary. In this model, the initial SOAP sender and the ultimate SOAP receiver are application modules within these organizations, and the security gateways are SOAP intermediaries. The model is illustrated below in Figure 3-4.

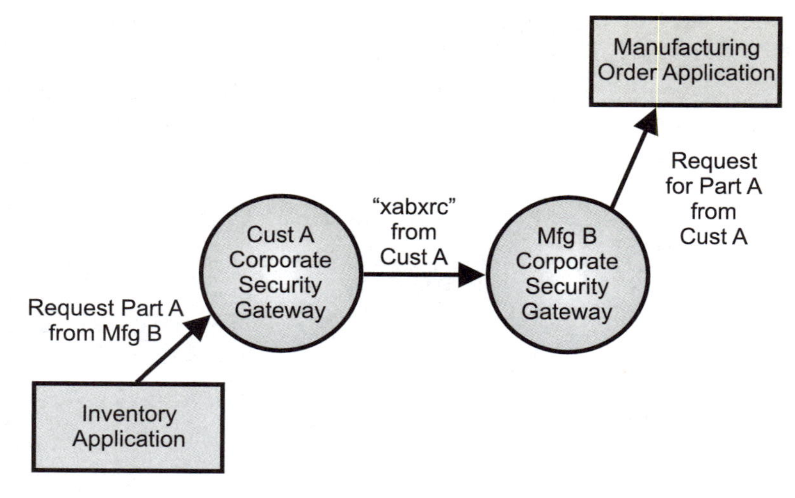

Figure 3-4. SOAP with security gateways.

3.2.1.4 SOAP with Attachments

An application can also send information as attachments associated with a SOAP message. This method can be used for sending large documents or binary content that do not require XML processing. The result is a reduction in XML parsing time and overhead. As a practical example, one could include a picture of a house associated with a real estate listing message. Another example is to include a document referenced in the primary SOAP envelope payload but not required for immediate processing. An application can associate attachments explicitly with a SOAP message (W3C, SOAP Messages with Attachments). This method utilizes a Multipart-MIME structure – a standard also used for e-mail – where one part contains the primary SOAP envelope and one or more parts contain attachments. Example 3-5 illustrates an example of a SOAP message with an image attachment.

```
Content-Type: multipart/related;
boundary="169.254.248.213.1.3652.1100568372.156.1"
Mime-Version: 1.0
--169.254.248.213.1.3652.1100568372.156.1
Content-Type: text/xml
<S11:Envelope xmlns:S11="..." xmlns:wsse="..." xmlns:wsu="..." xmlns:
ds="..." xmlns:xenc="...">
<S11:Header>
<wsse:Security>
<ds:Signature>
<ds:SignedInfo>
<ds:CanonicalizationMethod Algorithm='http://www.w3.org/2001/10/
xml-exc-c14n#'/>
<ds:SignatureMethod Algorithm='http://www.w3.org/2000/09/
xmldsig#rsa-sha1' />
<ds:Reference URI="cid:bar">
<ds:Transforms>
<ds:Transform Algorithm=3D"http://docs.oasis-open.org/wss/
```

```
04/XX/oasis-2004XX-wss-swa-profi le-1.0#Attachment-Content-Only-Transform"/>
</ds:Transforms>
<ds:DigestMethod Algorithm="http://www.w3.org/2000/09/
xmldsig#sha1"/>=20

<ds:DigestValue>Something</ds:DigestValue>
</ds:Reference>
</ds:SignedInfo>
<ds:SignatureValue>DeadBeef</ds:SignatureValue>
</ds:Signature>
</wsse:Security>
</S11:Header>
<S11:Body>
This is an image <image URI="cid:bar" />
</S11:Body>
</S11:Envelope>
--169.254.248.213.1.3652.1100568372.156.1
Content-Type: image/jpeg
Content-ID: <bar>
Content-Transfer-Encoding: base64
/9j/4RP+RXhpZgAASUkqAAgAAAAJAA8BAgAGAAAAegAAABABAgAVAAAAgAAAABIBAwA
...
qokzLkN931oKtc//2Q==
--169.254.248.213.1.3652.1100568372.156.1--
```

Example 3-5. SOAP message with attachment.

SOAP 1.2 also provides an alternative approach, where the XML processing tool automatically serializes binary-type element content as attachments, transparently creating a Multipart MIME structure on behalf of the application. This method is described in the MTOM specification. Regardless of the technique used, the receiver should be prepared to receive SOAP messages with additional content conveyed as attachments.

3.2.1.5 SOAP Transport

The SOAP protocol is defined at the application level of the ISO OSI protocol stack. The way that different services are provided by different protocol layers in a protocol stack is illustrated in the seven-layer networking model below (Figure 3-5). The lowest layers include the physical connection and the link layer. Examples are a local area network, a dial-up link, or a wireless network. This link layer can be quite complicated (including different message formats and control mechanisms), but it is simply used to transfer content or payload from one link endpoint to another. Built on top of this layer are additional protocols, such as TCP and IP, used to route payload from one network node to another in a network that can be extremely large (e.g., the Internet). As the core Web services messaging protocol, SOAP is designed to allow application modules to communicate with one another. It is transmitted using protocols lower in the protocol stack.

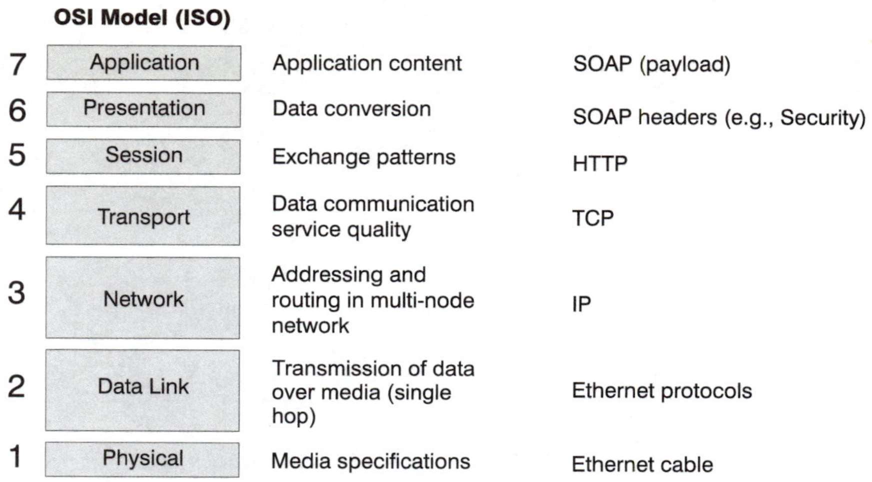

OSI Model (ISO)

7	Application	Application content	SOAP (payload)
6	Presentation	Data conversion	SOAP headers (e.g., Security)
5	Session	Exchange patterns	HTTP
4	Transport	Data communication service quality	TCP
3	Network	Addressing and routing in multi-node network	IP
2	Data Link	Transmission of data over media (single hop)	Ethernet protocols
1	Physical	Media specifications	Ethernet cable

Figure 3-5. OSI seven-layer model and SOAP/HTTP.

Understanding that protocols can be layered as a stack, where the protocol envelope of one layer is carried as the payload of the layer immediately below, makes it easier to understand what *SOAP over HTTP* means. The SOAP envelope is conveyed using the HTTP protocol just like any other application data. Similarly, HTTP messages are conveyed in TCP packets as TCP data.

SOAP messages are typically conveyed using HTTP or HTTPS (HTTP secured by SSL/ TLS). However, they can also be conveyed using other protocols, such as e-mail or Short Message Service (SMS) text messaging. A SOAP binding specification defines how the underlying protocol should convey SOAP messages and describes in detail how the two should work together.

3.3 Web Service Description

Web service description is an important part of Web services. It allows a Web service client to be configured to interact with a service without requiring custom software development or configuration. This is made possible by a machine-readable standard description of the service that automatically adapts the client to the specific service. The core specification in this area is the Web Services Description Language (WSDL). However, because WSDL does not cover all aspects of the interaction, additional description can be provided (e.g., information returned by a Liberty Alliance Discovery service).

3.3.1 WSDL Description

WSDL is an XML language used to describe the characteristics of Web service endpoints. A WSDL description defines the content or payload of a message (using XML Schema), the operations that the endpoint supports, and the specific aspects of the underlying transport protocol binding.

Let us consider the weather report example used earlier (www.webservicex.net). The schema definition shown in Example 3.6 can be included in the WSDL definition of the service, allowing a client to determine the payload format for the request and response. Note that the schema does not explain how to format the string. For example, the schema does not indicate that the string needs to be decoded. The WSDL description corresponding to the weather report example includes the following content (additional methods and `portTypes` have been removed to simplify the example):

```
<?xml version="1.0" encoding="utf-8"?>
<wsdl:definitions
    xmlns:http="http://schemas.xmlsoap.org/wsdl/http/"
    xmlns:soap="http://schemas.xmlsoap.org/wsdl/soap/"
    xmlns:s="http://www.w3.org/2001/XMLSchema"
    xmlns:soapenc="http://schemas.xmlsoap.org/soap/encoding/"
    xmlns:tns="http://www.webserviceX.NET"
    xmlns:tm="http://microsoft.com/wsdl/mime/textMatching/"
    xmlns:mime="http://schemas.xmlsoap.org/wsdl/mime/"
    targetNamespace="http://www.webserviceX.NET"
    xmlns:wsdl="http://schemas.xmlsoap.org/wsdl/" >
  <wsdl:types>
    --- schema, as shown above ---
  </wsdl:types>
  <wsdl:message name="GetWeatherSoapIn">
    <wsdl:part name="parameters" element="tns:GetWeather" />
  </wsdl:message>
  <wsdl:message name="GetWeatherSoapOut">
    <wsdl:part name="parameters" element="tns:GetWeatherResponse" />
  </wsdl:message>
  <!--  --- other methods not shown --- --!>
  <wsdl:portType name="GlobalWeatherSoap">
    <wsdl:operation name="GetWeather">
      <documentation xmlns="http://schemas.xmlsoap.org/wsdl/">Get weather
      report for all major cities around the world.
      </documentation>
      <wsdl:input message="tns:GetWeatherSoapIn" />
      <wsdl:output message="tns:GetWeatherSoapOut" />
    </wsdl:operation>
  </wsdl:portType>
  <wsdl:binding name="GlobalWeatherSoap" type="tns:GlobalWeatherSoap">
    <soap:binding transport="http://schemas.xmlsoap.org/soap/http"
                  style="document" />
    <wsdl:operation name="GetWeather">
      <soap:operation soapAction="http://www.webserviceX.NET/GetWeather"
                      style="document" />
      <wsdl:input>
        <soap:body use="literal" />
      </wsdl:input>
      <wsdl:output>
        <soap:body use="literal" />
```

```
      </wsdl:output>
    </wsdl:operation>
  </wsdl:binding>
  <wsdl:service name="GlobalWeather">
      <documentation xmlns="http://schemas.xmlsoap.org/wsdl/" />
      <wsdl:port name="GlobalWeatherSoap" binding="tns:GlobalWeatherSoap">
        <soap:address location="http://www.webservicex.net/globalweather.
                                asmx" />
      </wsdl:port>
      <wsdl:port name="GlobalWeatherHttpGet" binding="tns:
                                                      GlobalWeatherHttpGet">
        <http:address location="http://www.webservicex.net/globalweather.
                                asmx" />
      </wsdl:port>
      <wsdl:port name="GlobalWeatherHttpPost" binding="tns:

GlobalWeatherHttpPost">
        <http:address location="http://www.webservicex.net/globalweather.
                                asmx" />
      </wsdl:port>
  </wsdl:service>
</wsdl:definitions>
```

Example 3-6. WSDL description.

3.3.2 Technical Details

This section describes technical details related to the Web Services Definition Language (WSDL) as defined in WSDL 1.1. Submitted to the W3C by Ariba, IBM, and Microsoft, WSDL 1.1 is not a W3C Recommendation, but a W3C Note. It is widely adopted, so the explanations and examples used here correspond to WSDL 1.1.[2] At the time of writing, WSDL 2.0 is at W3C Working Draft status. Although it is not described here, the new version introduces numerous changes to the document structure. For the latest information, visit the W3C Web site.

A WSDL-document is divided into two parts: abstract and concrete. The abstract part defines XML-based message structures and data types of the message. In addition, this part establishes the message exchange sequences formally known as *operations*. The operation type is defined by the initiator of the message exchange: the client or the service provider. The operation is further specified by whether or not the recipient should reply to the message. Abstract and message-related issues are discussed in more detail below in the WSDL element-specific sections.

The concrete part defines the mechanics of the actual data communication. This includes the endpoints used to access services, and the selection of communication protocols supported by the services.

This separation of general message and data type definitions from the concrete mechanics makes it possible to reuse both parts. Changes in the transport method do not affect the message residing in the abstract part, and vice versa. The approach used in this section is top-down;

we start from the abstract data type definitions, and gradually work our way down to events taking place on the 'wire'.

A WSDL document contains six central elements: `types`, `message`, `portType`, `binding`, `port`, and `service`. The first three are abstract and the latter three concrete elements (Figure 3-6). Each element is covered in more detail in the following sections. Since their names do not clearly indicate their function, it is helpful to think of the elements as follows:

- The `types` element defines the XML types used as message payload. These are XML Schema definitions.
- The `message` element defines a logical one-way operation, such as a specific request message or a specific response message, in terms of the payload types.
- The `portType` element defines a message exchange, such as request-response, by collecting one or more `message` elements. This definition is abstract in the sense that the details of how the message exchange will operate are not defined, only the messages that are to participate in the exchange.
- The `binding` element specifies a transport for a message exchange. It defines the transport details for each message in the exchange, as well as general transport characteristics for the entire exchange. The purpose is to bind the transport protocol to the message exchange. One such transport protocol is HTTP.
- The `port` element is a real (concrete) endpoint for a message exchange (`portType`). It is defined in terms of the transport protocol bound to the exchange (`binding`). For example, when HTTP is used, a port element would specify the endpoint URL.
- The `service` element is a collection of endpoints that offer meaningful, related operations associated with a service offering. Thus, a `service` element defines the various exchange patterns in concrete terms (a `port` for each `portType`).

The WSDL document is described in more detail below by using an example of a simple Internet bookstore. The features used as examples enable clients, for example, to check for the availability of book titles.

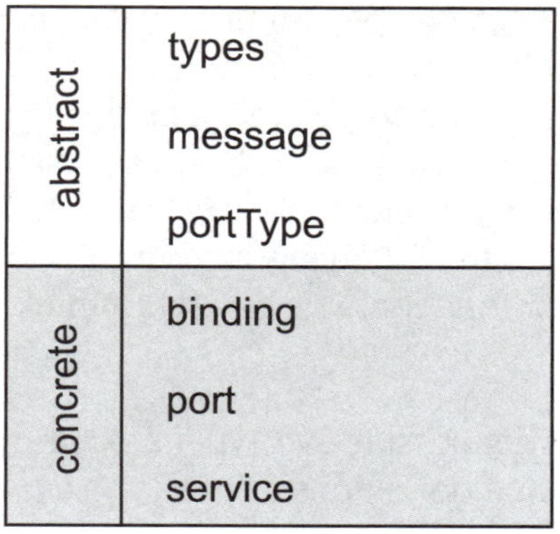

abstract	types message portType
concrete	binding port service

Figure 3-6. WSDL document elements.

3.3.3 WSDL Document Structure

The root element of all WSDL documents is called `definitions`.[3] All namespace definitions are placed within the root element (Example 3-7). This is the root XML document that contains the WSDL definitions.

```
<definitions xmlns=”http://schemas.xmlsoap.org/wsdl/” xmlns:soap=”http://
schemas.xmlsoap.org/wsdl/soap/”
xmlns:http=”http://schemas.xmlsoap.org/wsdl/http/”
xmlns:xs=”http://www.w3.org/2001/XMLSchema/” xmlns:tns=”http://ji.example.
com/examplewsdl/”
targetNamespace=” http://ji.example.com/examplewsdl”/>
.
.
.
</definitions>
```

Example 3-7. Root element of a WSDL document.

3.3.3.1 Types – XML Schema definitions for Messages

The `types` element defines the data types used in the WSDL document and by the messages during communication with the service provider.[4] XML Schema is considered to be the essential WSDL type definition language, although the WSDL specification includes expansion methods to cater for any future data type definition languages.

As Example 3-8 below demonstrates, the types element contains normal XML Schema data. (For a detailed description of XML Schema, see chapter 2.)

```
<types>
 <xs:schema targetNamespace=”http://ji.example.com/examplewsdl”>

  <xs:element name=”qtyCheck”>
   <xs:complexType>
    <xs:all>
     <xs:element name=”nameOfBook” type=”xs:string” />
```

```
    </xs:all>
   </xs:complexType>
 </xs:element>

 <xs:element name="qtyReply">
  <xs:complexType>
   <xs:sequence>
    <xs:element name="qtyOfBook" type="xs:integer"/>
   </xs:sequence>
  </xs:complexType>
 </xs:element>

 </xs:schema>
</types>
```

Example 3-8. The `types` *element.*

All data type definitions need not be defined in the types element, since it is possible to also reference built-in XSD types from within a message element.

3.3.3.2 Message – One-way Message definition

The `message` element defines an entity that contains abstract definitions of the data types used in the message.[5] The element defines a one-way message in terms of the XML data. Other WSDL components build a message exchange pattern using message elements and bind these patterns to transport protocol information.

A separate `message` element is defined for each service message, where one message is used, for example, for a query, and other for a reply. An individual `message` element could be defined for a feature that queries for the quantity of a certain title in an Internet bookstore, and another for the reply returning the number of available items. Each `message` element must have a unique name within the WSDL document.

Each `part` element in a `message` element defines message-specific data types by referencing the data types defined in the types element or by using a built-in XML Schema `simpleType` or `complexType` element.

Let us use a simple Web bookstore as an example. The store's book availability feature uses two messages: a query that sends the name of the requested title to the store, and a reply that contains the number of available books. The corresponding data types are defined in the `types` element. Note that each `message` element must have a unique name within the WSDL document, and each `part` must also be uniquely named within the `message` element.

```
<message name="QtyCheckInput">
<part name="pt" element="tns:qtyCheck"/>
</message>
```

```
<message name="QtyCheckOutput">
<part name="pt" element="tns:qtyReply"/>
</message>
```

Example 3-9. The `message` element.

The above example defines the `part` types by referencing the `qtyCheck` and `qtyReply` elements defined in the XML Schema embedded in the types element.

The number of `part` elements depends on the number of variables in each service message. A `message` element can contain several parameters, where each parameter (i.e., `part` element) must have a unique name within the `message` element. The following example illustrates a `message` element of a simplified book purchase feature. The service uses the name of the book and the quantity of ordered titles as parameters, and the example `message` element defines the corresponding data types by using XML Schema data types. Note that all types are defined within the `message` element – the `types` element is not needed at all for this message.

```
<message name="BuyBook">
        <part name="bookName" type="xsd:string"/>
        <part name="quantity" type="xsd:integer"/>
</message>
```

Example 3-10. The `part` element within the `message` element.

3.3.3.3 PortType – message exchange pattern

The `portType` element defines an abstract entity that collects all input and output messages related to a single service feature.[6] A service feature here refers to a logical transaction carried out to achieve the desired result, regardless of whether this takes one (notification) or two (query-reply) messages. The number of messages and the initiator of the transaction (direction of message flow) are defined by an `operation` element, which determines the chosen operation type (i.e., *transmission primitive*) out of four possible alternatives:

- **one-way** – The client sends a message to a service provider, but the service provider does not send a reply. This means that the `operation` element contains an `input` element only.
- **request-response** – The client sends a message to a service provider. The service provider receives the message and sends a response message to the client. Both message directions require an element of their own; thus the `operation` element contains one `input` and one `output` element.

- **solicit-response** – The service provider initiates the communication by sending the a message to the client. The service provider expects to receive a response to the original message. As in the previous example, the `operation` element contains one `input` and one `output` element.
- **notification** – The service provider sends a message to the client, but does not expect a response. One-way messaging requires just one element – thus the `operation` element only contains an `output` element.

In the bookstore example, the title availability check feature uses two messages: `qtyCheckInput` for input and `qtyCheckOutput` for output. (The messages are defined in the respective `message` elements.) The `portType` element has a unique name within the WSDL document, should there be more than one `portType` elements.

```
<portType name="QtyCheckPortType">
<operation name="CheckQuantity">
<input message="tns:qtyCheckInput"/>
<output message="tns:qtyCheckOutput"/>
</operation>
</portType>
```

Example 3-11. The `portType` element.

The `operation` elements also allow the definition of optional `fault` messages, which are used in case of certain error situations.

3.3.3.4 Binding – define transport for a message exchange pattern

From a layered, architectural point of view, the `binding` element resides between the abstract `portType` element and the actual service network address.[7] Logically, it follows that the `binding` element defines the transport protocol attributes and message formats used to relay messages to the network address.

The `binding` specification allows multiple transport methods. The use of SOAP 1.1 with HTTP is common, and therefore we use SOAP1.1/HTTP binding in our example.[8]

In the example, the binding element defines a SOAP/HTTP transport method for the `portType` element named `QtyCheckPortType`. The first two attributes (`name` and `type`) of the element are rather self-explanatory; the `name` attribute gives the element a unique `name` and the `type` attribute references the `portType` element that relies on this element for transport services. This binds the message exchange pattern (`portType`) to the transport protocol information contained in the binding element. The actual messaging type, transport protocol, and encoding of the message data are defined with the soap:binding, soap:operation, and soap:body elements.

The soap:binding element announces that the binding uses SOAP messaging, and that the form of the SOAP messaging is either Remote Procedure Call (RPC) or document-based.[9] In addition, this element defines the transport protocol. The element has two attributes:

- The style attribute defines the SOAP messaging model to be used in the binding. The alternatives are RPC and document-based, and the respective attribute values are rpc and document.
- The transport attribute defines the transport protocol with a URI. For example, the HTTP protocol that the bookstore example uses is defined with the URI <http://schemas.xmlsoap.org/soap/http>. Other alternatives include, for example, HTTPS and SMTP.

The soap:operation element embedded within an operation element contains information about a single operation. The element has two attributes:

- The soapAction attribute defines the character string placed in the SOAPAction HTTP header.[10]
- The style attribute selects the SOAP messaging model for the operation. This attribute overrides the style attribute of the soap:binding element. As with the soap:binding element, the possible attribute values are rpc and document.

The soap:body element declares how the message parts are placed in a SOAP body element and how the parts are to be transmitted in a message. If the literal option is specified, then each message is sent on the transport protocol by serializing the XML InfoSet representation for the XML type (or as stated in the WSDL Note, "each part references a concrete schema").[11] This is a commonly used form in Web service deployment. If an encoded form is selected, then the output of an encoding (of the type defined for the message associated with the operation) is placed in the message stream. The encoding is specified by the encodingStyle attribute.

The literal method does not necessarily involve any additional attributes or methods. The soap:body element is simply: <soap:body use="literal"/>. One element is defined for input and another for output messages. The following example illustrates the binding used in the bookstore example:

```
<binding name="QtyCheckSOAPBinding" type="tns:QtyCheckPortType">
<soap:binding transport=http://schemas.xmlsoap.org/soap/http
style="document"/>
<operation name="CheckQuantity">
<soap:operation soapAction=http://ji.example.com/ style="document"/>
<input>
<soap:body use="literal"/>
</input>
<output>
<soap:body use="literal"/>
</output>
</operation>
</binding>
```

Example 3-12. The `binding` element using literal binding method.

3.3.3.5 Port – endpoint defined for MEP bound to transport

Each `port` element links a binding with a protocol-specific network address.[12]

The functionality and the contents of a `port` element are rather simple. The element contains two attributes and one element:

- `name` – This attribute names the element uniquely with respect to other port elements in the document.
- `binding` – This attribute refers to the binding for which this `port` element provides the network address.
- `soap:address` – This element defines a URL for the protocol in question.

In the following example, we define a SOAP endpoint for the binding presented in the previous section.

```
<port name="QtyCheckPort" binding="tns:QtyCheckSOAPBinding">
<soap:address location=http://j.example.com/qtycheck/>
</port>
```

Example 3-13. The `port` element.

3.3.3.6 Service – Collection of related MEP endpoints

A `service` element simply collects a group of `port` elements under the same element, as illustrated in the following example:[13]

```
<service name="QuantityCheck">
<port name="QtyCheckPort" binding="tns:QtyCheckSOAPBinding">
<soap:address location=http://ji.example.com/qtycheck/>
</port>

</service>
```

Example 3-14. The `service` element.

3.4 Web Service Discovery

Before a Web service can be invoked, it needs to be discovered. Discovery refers to the process of obtaining the information needed to use the service, including interface and protocol information (such as that included in the WSDL document), and XML Schema information on the message content (this can also be included in the WSDL document). It may also be necessary for the client to include additional information in the SOAP message header for the service to accept the request. For example, security tokens can be used to indicate that the client has established its identity with a trusted authority, or that the client is authorized to make the request. Other information needed to use the service, such as general policy and processing requirements, can also be discovered.

Discovery can take place in several ways. A requester can discover a Web service, for example, in the following ways:

- By being told about the service out of band. Examples include obtaining the information from a service provider by e-mail, or by being dynamically informed about the service during an HTTP transaction (for example, by using the Liberty Reverse HTTP Binding for SOAP Specification).
- By visiting a well-known location. Knowledge of this location can be, for example, shared out of band, discovered on a Web site, or shared as *metadata*.
- By using a directory, such as a UDDI directory.
- By using an identity-based discovery service.

One example of out-of-band discovery is enterprises exchanging WSDL information as part of establishing a partner agreement. Another example is an informal interoperability event where participants exchange interface information by e-mail.

Use of well-known locations is common. A reference to the WSDL description of the weather service example discussed earlier can be located on the Web site of a company that offers a variety of Web services. This site could be discovered by using a Web search engine such as Google, or it could be found in a reference such as a magazine article describing useful Web services.

Another approach is to systematically register information about services in a directory service, such as a UDDI directory. A private UDDI directory can be used within an enterprise context, or shared within the enterprise and with partners (but not with the public to avoid potential denial of service attacks or other problems resulting from wide sharing of private information). However, a UDDI directory can also be public. In both cases, the used mechanism is the same. A UDDI directory provides a variety of information, allowing the retrieval of business information through searches on company name, product classification, or other criteria. The retrieved content can be the WSDL definition of services or other information related to the service.

Different discovery mechanisms are available. For services that are related to the identity of the participant, a system such as the Liberty Discovery Service may be appropriate, since this service enables the discovery of only those services that are appropriate to the identity of the requester. Other scenarios may require the general ability to locate services based on service type, industry or other criteria, calling for a more general directory, such as the UDDI service.

The remainder of this chapter focuses on the features of a UDDI registry. The another central discovery solution, i.e., the Liberty Discovery Service, is covered in more detail in chapter 5.

3.4.1 UDDI

The Universal Description, Discovery, and Integration (UDDI) specification defines methods for storing, locating, and managing service information related to Web Services.[14] The information is comprised of the general characteristics of service providers and their services, as well as technical information needed to access the services. UDDI is similar to white pages (listing of telephone numbers indexed by name) and yellow pages (listing of businesses indexed by service type). It also contains a set of information organized by industry. Application developers can use this information to locate services and information needed to create client applications that are compatible with the provided services. UDDI itself is a Web service, and its functionalities are based on Web services technology: the transport services are implemented using HTTP, the data structures are based on XML, and the messaging is based on SOAP messages.

Web service providers can use UDDI to register their Web service offerings. This allows Web service consumers to locate Web services according to their interests. UDDI enables services to be indexed according to a variety of classifications. Categorizing service providers and services by classification and identifier attributes[15] is particularly useful for two reasons: it allows Web service consumers to discover one or more similar services, and it allows Web service providers to announce their services in a meaningful manner. For example, if a service provider announces that it offers a service in the category called *fish farmers*, then Web service consumers can locate the service provider without prior information. The UDDI classification metadata system is tied to traditional industry classification systems, which allows it to leverage existing classifications and facilitate adoption. UDDI supports multiple classification systems, including the United Nations Standard Products and Services Code System (UNSPSC) and the North American Industry Classification System (NAICS).[16] The identifier feature can be used to identify the service provider uniquely by using a third-party identifier system (e.g., by listing the company's Global Location Number).[17] Another possible classification criterion is related to the service provider's geographical location.

The UDDI technical service information can specify data related to the service interface through which client applications are able to connect to and use the services. This information may involve re-using the WSDL description mechanism. The technical aspects of UDDI are covered in more detail below.

In this section, the description of UDDI is divided into two parts: registry and programming APIs. The aim is to provide an overview of the role of UDDI in a SOA. While a detailed description of the internal data structures is beyond the scope of this book, references to more detailed information are provided where necessary.

3.4.1.1 UDDI Registry

The database that stores service information is called a UDDI Registry. The role and capabilities of a UDDI registry have evolved from simple data storage to complex registries including capabilities important for modern enterprises, such as distributed computing, replication, and a policy system used to control registry operation. It is important to note that a UDDI registry is a logical component that can be implemented over one or more physical computers to enable enterprise features.

3.4.1.2 UDDI Registry Categorization

UDDI registries can be divided into three different domains according to visibility and data replication characteristics: public, affiliated (i.e., semi-private), and private.[18]

1. Public UDDI registries can be accessed freely by any client. Information stored in the registries can also be replicated by other registries. The UDDI Business Registry (UBR) is worth special attention. The UBR is a public UDDI registry, which is composed of a set of nodes hosted by IBM, Microsoft, SAP, and NTT Com (for an up-to-date list, visit <http://uddi.org/find>).

2. Private registries are specific to an organization and cannot be accessed from outside the organization. This allows larger organizations to manage the discovery of internal Web services while avoiding the security risks associated with allowing access to outsiders. Private registries can be shared within an extended enterprise (for example, an organization and its partners).

3. Affiliated registries can be used by organizations, which want to manage and control their own private registries, but also want to share them with selected organizations. By agreeing to a common indexing format, the organizations can link their registries and enable searches across affiliated registries. For example, a wholesaler and several retailers could affiliate their registries. Data replication and subscription are controlled by and limited to the affiliated organizations. This type of registry allows business partners to share services within a restricted environment while managing and controlling their own registries.

3.4.1.3 Data Models

In a UDDI registry, data is stored in XML format, and data types are defined using XML Schema. The information is stored in the following six element types:[19]

- A `businessEntity` contains general information about the company or organization offering the Web Services.
- A `businessService` element contains general information about a single service offering. From a technical point of view, even though it offers a single composite Web service interface, the service can be comprised of several individual Web services. A `businessService` is provided by a `businessEntity`.
- A `bindingTemplate` element contains technical information (particularly endpoint information) required to use a service described in the `businessService` element.
- A `tModel` element contains description information related to the `businessService`. A `tModel` is separated from the hierarchy of the above three entities into a data structure of its own, and therefore can be shared by multiple services. `tModel` description information can be defined using WSDL.
- A `publisherAssertion` element can be used to describe the cooperative relationship between two organizations, each specified in a `businessEntity`.

- A `subscription` element allows parties to register to be notified about changes in UDDI data (see section 3.4.2.3 below).

3.4.2 UDDI Application Programming Interfaces

The UDDI Programmers APIs provide features for publishing information, searching and modifying data, controlling UDDI registries, and managing collaboration.[20]

3.4.2.1 Inquiry API

The Inquiry API provides functions for searching service provider and service information from registries.[21] The API provides three different inquiry patterns: browse, drill-down, and invocation.

- The *browse pattern* is used to fetch general information using general search parameters. The resulting information can later be used to obtain more detailed information (by using the drill-down pattern).
- The *drill-down pattern* uses an ID key as an argument and fetches the corresponding `businessDetail` and `businessEntity` information.
- The *invocation pattern* is used to fetch information that enables the actual use of the Web service. This information is obtained from the `bindingTemplate`.

3.4.2.2 Publication API

The publication API provides features for adding, modifying, and removing registry data.[22] The data can be managed in two ways: by using a programmatic method to communicate directly with the registry, or by using a Web user interface.

3.4.2.3 Subscription API

Affiliated registries can use a subscription system to receive notifications about changes to data stored in other registries.[23] This ensures that up-to-date information is available to all affiliated parties. The notification system can function in a synchronous or asynchronous manner. With the synchronous method, a registry that wants to check for possible changes made to another registry sends out a request asking to be updated. With the asynchronous method, a registry sends intermittent updates to all parties that have active subscriptions with the registry.

3.5 Summary

This chapter has introduced the core concepts related to Web services: messaging, description, and discovery. The following chapters outline the more advanced areas of addressing, policy, identity, and security. Appendix A provides more detailed information on the standards organizations referred to in this chapter.

Chapter 4: Agreement

In this chapter, we introduce two new concepts related to Web services: addressing and policies. Web services addressing improves SOAP messaging by separating SOAP messaging from the transport protocol used to convey the messages. The policy system (in effect, the actual policy definition languages) is used to create formal rules guiding, for example, security requirements.

4.1 Web Services Addressing

Web Services Addressing (WS-Addressing) defines mechanisms used to target SOAP messages to SOAP processors in complex application scenarios. It is a critical component of the Web services messaging layer, and in many ways, it should have been part of the SOAP 1.2 specification. Before WS-Addressing, the Web services architecture conflated the messaging layer and the transport layer. The separation was not clean, as much of the SOAP addressing information was defined in the transfer protocol, typically HTTP. WS-Addressing places all SOAP information at the SOAP messaging layer, thus providing a more sound architecture for applications.

One of the principal advantages of this new layering is that it enables the long-held promise of transport protocol independence. Since all the addressing information is available as SOAP header information, SOAP nodes (including intermediaries) are able to meet application requirements by drawing from a heterogeneous mix of transport protocols.

4.2 Background

Web services are defined with a family of specifications each targeted to define a different aspect of the functionality of Web services. The layer providing basic message exchange capability is called the XML Protocol layer, and the applicable specification for this layer is SOAP. In the various descriptions of the Web services architecture, this layer is also known as the messaging layer.

While the SOAP specification dominates this layer of the architecture, there are emerging specifications for expanding Web services core messaging capabilities beyond SOAP and for filling out certain technical niches. These niches exist where SOAP is either underspecified or the new capability is outside the scope of SOAP.

This description of WS-Addressing focuses on key specification work that involves realigning the layering between SOAP and the underlying transfer protocol. This is the concept of addressing Web service messages, and the applicable standard is WS-Addressing.

4.3 Purpose of WS-Addressing

At a high level, the purpose of WS-Addressing is to enable SOAP messages to flow over multiple protocols or distinct sessions of the same protocol. WS-Addressing also allows a Web service designer to create more flexible messaging patterns, if they are needed in order to implement an application. One of the most important patterns – especially for mobile Web services – is asynchronous messaging. Asynchronous messaging systems and systems requiring application-unique messaging patterns can now be deployed for Web services by using a standardized approach that utilizes the features provided by WS-Addressing. Note that at the time of writing, the WS-Addressing standard does not specifically define how to implement asynchronous SOAP messaging. It does provide, however, the essential addressing features that a Web service designer can exploit to create an asynchronous service.

SOAP left many important details to the transport layer, such as specifying where a message is going, how to respond, how to multicast, where errors should go, and so on. It also left many important dispatch details to the transport and the platform, such as how to get a message to the correct machine, to the correct service, through a server farm, to the right WSDL interface or port, and to the correct service handler code. For the transport issues, all the heavy lifting has been done by HTTP while many other protocols have been shut out. For the dispatch issues, each vendor has created its own conventions that service deployers must be aware of. The current situation is poor in terms of interoperability or allowing innovative and more complex applications and services to appear in the market.

WS-Addressing is intended to help solve this dilemma. WS-Addressing will become a key component of the emergent Web services architecture. Many specifications and standardization efforts will use the layering provided by WS-Addressing and will be able to concentrate better on the semantics of their specification rather than the common services of message addressing in a transport and platform-agnostic distributed environment.

4.4 Mobile Application Example

Before getting into the details of what WS-Addressing exactly is, how it is used, and why it is important, let us begin with an example. Our example application provides mobile enterprise calendar scheduling. The scenario is to create a meeting event, initiated by mobile SOAP client. The request from the client is sent to a SOAP-aware meeting scheduling service. Described later in more detail, this service will be long-lived and hence asynchronous. Due to cellular data characteristics, reaching a mobile phone is problematic using HTTP. Therefore, the SOAP response is sent to the mobile using the SMS protocol. Since many services cannot support multiple transfer protocols, we designed the example by using a gateway intermediary to move the SOAP response from the HTTP protocol to the SMS protocol.

```
<S:Envelope xmlns:S="http://www.w3.org/2003/05/soap-envelope"
            xmlns:wsa="http://www.w3.org/2005/08/addressing"
            xmlns:mobile-sms=http://www.nokia.com/mobile/sms/addressing
            xmlns:mobile-apps=http://www.nokia.com/mobile/apps
            xmlns:nokian=http://www.nokia.com/employees
            xmlns:companyx=http://www.companyx.com/employees>
  <S:Header>
    <wsa:MessageID>
      http://178155512.mobile.nokia.com/ids/1234567890abcdefghijklmn
    </wsa:MessageID>
    <wsa:ReplyTo>
```

```
      <wsa:EndpointReference>
       <wsa:Address>http://sms-gateway.nokia.com</wsa:Address>
       <wsa:ReferenceParameters>
          <mobile-sms:terminal-id
            S:role="http://www.w3.org/2003/05/soap-envelope/role/next"
            S:mustUnderstand="true">
            +17815551212
          </mobile-sms:terminal-id>
          <mobile-sms:app-port
            S:role="http://www.w3.org/2003/05/soap-envelope/role/next"
            S:mustUnderstand="true">
            1025
         </mobile-sms:app-port>
       </wsa:ReferenceParameters>
      </wsa:EndpointReference>
     </wsa:ReplyTo>
     <wsa:FaultTo>
      <wsa:EndpointReference>
       <wsa:Address>http://17815551212.mobile.nokia.com/calendar/log</wsa:
Address>
      </wsa:EndpointReference>
     </wsa:FaultTo>
     <wsa:To>http://calendar.nokia.com</wsa:To>
     <wsa:Action>http://calendar.nokia.com/new-event-request</wsa:Action>
   </S:Header>
   <S:Body>
     <mobile-apps:meeting-scheduler>
       <mobile-apps:confirmed-first-avail-free-slot>
         <nokian:employee-id>10073708</nokian:employee-id>
         <nokian:employee-id>10073709</nokian:employee-id>
         <companyx:employeenum>987-654-321</companyx:employeenum>
       </mobile-apps:confirmed-first-avail-free-slot>
     </mobile-apps:meeting-scheduler>
   </S:Body>
</:Envelope>
```

Example 4-1. The SOAP request message used to book a meeting.

The salient, operational constructs of the example application related to the above SOAP request message are:

- A set of SOAP headers (in the `wsa` namespace) that correspond to WS-Addressing features
- Use of `MessageID`, a message-identifying element, to correlate messages. In the example, this message identifier was generated by the mobile phone that created the SOAP request.
- Use of the `ReplyTo` element in the reply message to achieve transport independence. In the example application, the value of this element targets the gateway intermediary that provides the bridge between HTTP and SMS transports.
- Use of `ReferenceParameters`, extensibility points that enable finer-grained message addressing. They can be attached to any addressing element, such as `To`, `ReplyTo`, `FaultTo`, and `From`. In the example application, we use reference parameters to define

the elements needed by the gateway in order to address the mobile terminal, the ultimate receiver of the SOAP response.

- The separation of reply messages and faults using the `FaultTo` element.
- Use of the `To` element to place the address of the target SOAP processor at the SOAP layer, rather than at the lower transport layer.
- Surfacing of the `Action` header to the SOAP layer. This header identifies the semantics of the operation and provides a means for dispatching the SOAP message to the correct handler application. It also provides the association to a WSDL description of the service.
- Refining the requested service via the SOAP body, for example, by locating the first available free time slot for all meeting participants and ensuring that all participants confirm the slot selection. The semantics of the example application provide the rationale for a long-running transaction. Thus, once the calendar service has received the request, it must locate a suitable time slot and receive a confirmation from each participant on the meeting time. This confirmation forces the service to be lengthy and asynchronous. (To keep the example simple, the confirmation messaging round is not shown.)
- If a SOAP-layer fault occurs at the service provider's end, the SOAP response would follow a different path in the example. Instead of the routing back to the mobile terminal via the SMS gateway, the SOAP fault response would go back directly to the mobile terminal via the existing HTTP session. This follows from value of the `FaultTo` header matching that of the originator of the request message: `<http://17815551212.mobile.nokia.com/calendar/log>`. An alternative value of `ReplyTo` can be used for the same effect – to transmit a response to the originator. This works for both `ReplyTo` and `FaultTo` headers. We discuss this predefined value `<http://www.w3.org/2005/08/addressing/anonymous>` in more detail later.

The following example illustrates a successful use case, in which a response SOAP message is sent to the SMS. The response is transmitted using a new HTTP session established between the calendar service and the SMS gateway SOAP intermediary.

```
<S:Envelope xmlns:S...>
  <S:Header>
    <wsa:MessageID>
       http://cal-service.nokia.com/ids/0987654321nmlkjihgfedcba
    </wsa:MessageID>
    <wsa:RelatesTo>
       http://178155512.mobile.nokia.com/ids/1234567890abcdefghijklmn
    </wsa:RelatesTo>
    <wsa:To>http://sms-gateway.nokia.com</wsa:To>
    <mobile-sms:terminal-id
        S:role=http://www.w3.org/2003/05/soap-envelope/role/next
        S:mustUnderstand="true"
        wsa:IsReferenceParameter='true'>
        +17815551212
    </mobile-sms:terminal-id>
    <mobile-sms:app-port
        S:role="http://www.w3.org/2003/05/soap-envelope/role/next"
        S:mustUnderstand="true"
        wsa:IsReferenceParameter='true'>
        1025
    </mobile-sms:app-port>
    <wsa:Action>http://calendar.nokia.com/new-event-callback</wsa:Action>
  </S:Header>
  <S:Body>
    <mobile-apps:meeting-scheduler>
      <mobile-apps:confirmed-new-event>
        <!--calendar event is here --->
      </mobile-apps:confirmed-new-event>
    </mobile-apps:meeting-scheduler>
  </S:Body>
```

Example 4-2. The SOAP response to the meeting request from the calendar service to the SMS gateway.

The salient, operational constructs of the example application related to the above SOAP response message are:

- A unique `MessageID` has been created for this SOAP message. This message identifier was generated by the calendar service.
- A correlation element, `RelatesTo`, associates multiple messages as part of the same service provision.
- The `To` address header of the SOAP response takes the same value as the `ReplyTo` header of the SOAP request.
- The reference parameters of the `ReplyTo` header of the SOAP request become SOAP header child elements. Thus, in our example, the terminal-specific addressing elements are surfaced for processing by the gateway intermediary.

As the final messaging stage of the successful use case, the gateway intermediary transmits the following SOAP message through the cellular network to the target mobile terminal using the SMS protocol for SOAP message transport.

```
<S:Envelope xmlns:S…>
  <S:Header>
    <wsa:MessageID>
      http://cal-service.nokia.com/ids/0987654321nmlkjihgfedcba
    </wsa:MessageID>
    <wsa:RelatesTo>
      http://78155512.mobile.nokia.com/ids/1234567890abcdefghijklmn
    </wsa:RelatesTo>
    <wsa:To>sms:+17815551212:1025</wsa:To>
    <wsa:Action>http://calendar.nokia.com/new-event-callback</wsa:Action>
  </S:Header>
  <S:Body>
    <mobile-apps:meeting-scheduler>
      <mobile-apps:confirmed-new-event>
        <!--calendar event is here --->
      </mobile-apps:confirmed-new-event>
    </mobile-apps:meeting-scheduler>
  </S:Body>
```

Example 4-3. The SOAP message from the SMS gateway to the terminal.

The gateway intermediary reads the SOAP headers related to the SMS terminal ID and application port (app-port), and uses them to replace the value of the To header with a single URI (an SMS identifier). This identifier has the form `sms:<terminal-id>:<port>`. Although intermediary behavior is underspecified in WS-Addressing, this is the semantics to be expected from an intermediary whose role is to be a protocol gateway. As the gateway is an intermediary and not an ultimate receiver, the message remains a SOAP response and the message identifier remains unchanged.

The SOAP response is dispatched to the mobile terminal using the action header. The SOAP client application receiver handles the message asynchronously by receiving the message via a different transport protocol than the one used to convey the SOAP request. This is a so-called callback scenario: the response is handled on a different thread than the thread that made the initial HTTP request. The receiving mobile application may use the correlation data (the value of the `RelatesTo` header) to access the application context that made the initial request.

This example incorporates many of the technical features of WS-Addressing. Perhaps more importantly, the above example illustrates much of the rationale and importance of WS-Addressing to the Web services architecture. For complex applications, WS-Addressing provides critical core elements for achieving interoperability. The types of applications that benefit from WS-Addressing are those that need to support multiple transfer or transport protocols, require fine-grained dispatch semantics at the application level, include long-running services or asynchronous messaging, or some combination of these system features.

4.5 WS-Addressing Core

WS-Addressing is fairly simple and straightforward. However, familiarity with XML specifications helps in understanding the specification. The Web Services Addressing 1.0 – Core specification defines two main entities: *endpoint references (EPRs)* and *message addressing properties*. The specification also defines processing rules for constructing a response message when a request message with WS-Addressing headers is received. We discuss these rules in the context of message addressing properties. Note that the SOAP header elements and attributes are actually described in the SOAP Binding specification of WS-Addressing. The Core specification defines the new entities only at the abstract level of the XML Infoset.

EPRs and message addressing properties are closely related, and do not represent very different or distinct capabilities. They follow a strong software design tradition of reuse, modularity, and composability. EPRs provide a core layer of abstraction and message addressing properties are composed of EPRs that define each new kind of address in the specification. For instance, the `ReplyTo` concept has at its core an EPR, as do the other two primary WS-Addressing concepts we saw in the example: the `To` and `FaultTo` headers. Thus, as in all proper software and information design, WS-Addressing is a layered specification that relies on the principals of reuse, modularity, and composability to achieve a succinct and powerful specification.

4.5.1 Endpoint References

An EPR forms the basic building block in WS-Addressing. As stated above, EPRs are used to define various kinds of endpoints defined in the message addressing properties. The Core specification mentions EPRs to satisfy some distinct usage scenarios: dynamic generation and customization of endpoint descriptions, referencing of endpoints used for stateful interactions, and expression of refined endpoint information in tightly-coupled environments with well-known policy assumptions.

Hence, EPRs are designed for two main purposes: finer-grained SOAP-layer dispatching, and stateful interactions. The following sections describe the elements of SOAP dispatching and stateful services, and the functioning of EPRs.

4.5.1.1 Dispatch

Dispatching refers to the necessary process of moving a message from its arrival on a device to its application processor. The application processor is typically executable code that must be started or invoked. Mapping the arrived message to the correct executable code with the correct input parameters is the difficult part of dispatching.

SOAP dispatch is typically comprised of two steps. First, the message must arrive in the correct device. This is properly the task of the underlying transport when supplied with the correct URL. The URL scheme identifies the transfer protocol: HTTP, mailto, ftp, etc. The transport's task is to get the message to the right device, such as calendar.nokia.com. Complexities such as server farms and redirects are handled by the transfer protocol.

The second step is to identify the right service on the device, since a single DNS-addressable device can host multiple SOAP services. Before WS-Addressing, the service was identified via the URL given to the transport. Hence, the URL `<http://calendar.nokia.com/scheduler>` would indicate that the service to run is the scheduler service. The dispatching software would have to know something about the transport layer to recognize the service identifier in the URL handed to the transport layer for moving the message to the correct device. Surfacing the service identity to the SOAP layer is one of the features of WS-Addressing. Decoupling the executing device from the service name is a function of EPRs.

An EPR must be defined to allow the correct service to be identified at the SOAP layer. This allows the message to be dispatched without undue assumptions being made about the deployment platform.

When a message it is received it must be passed to the correct application processor, and the application processor must determine the correct operation to perform. The EPR is used to determine the correct application processor, while the message addressing properties and SOAP `Action` header are used to invoke the correct operation. This separation of endpoint and operation makes sense in that services are the level of granularity of endpoints, while operations are the public methods they offer. This final stage of SOAP dispatch involving the `Action` header is discussed in section 4.5.2 describing message addressing properties.

To summarize, the purpose of EPRs is to allow the transfer protocol to get the message to the right device, and after that, layer the service and operation information in the SOAP message in such a way that the SOAP-layer platform can inspect only the SOAP message in order to perform the proper dispatching. Thus, WS-Addressing enables a standardized model for information deployment via EPRs and the `Action` header instead of relying on Servlet, ASM, or some other deployment technology.

4.5.1.2 Stateful Services and Messaging

Support for stateful interactions is a requirement of WS-Addressing articulated as one of the usage scenarios in the context of EPRs. Note that the Web was initially designed for stateless services. However, demand for a service state to be passed as a token between client and server in browser-based applications has led to the creation of Web cookies, encoding of URL parameters, and HTTP POST data. In similar fashion, Web services need to support both stateful services and stateful interactions.

A quick set of definitions should help understand stateful services and messaging. In a stateful Web service application, an application state must be created and maintained while the service invoker and the service provider exchange messages in the scope of provisioning a single service.

There are two ways to implement a stateful Web service application. The first is to use a stateful service, in which case the application state resides on the server and some kind of state identifier is embedded in the messages. The second way is to use stateful interactions or messages, in which the full application state is encoded in the message in some fashion and passed back and forth. What can be particularly confusing is that stateless service is commonly used to refer to the form of interaction where the full state is resident in the message and not on the server – just as in a stateful message. EPRs can be useful in both forms of interaction. For stateful interactions, the state can be encoded using reference parameters. For stateful services, reference parameters can be used to enable the SOAP-layer processor to perform fine-grained dispatches to the service endpoint managing the application state.

As applications become increasingly complex, application designers have to weigh the costs and benefits of each of these interaction models with respect to their application's functional and non-functional requirements. Stateless services are best suited when a single service has a large number of users, the extent of state exchanged is relatively small, and the number of interactions per service is low. Note that most enterprise integration, B2B, and even retail shopping cart applications do not meet the criteria of a stateless service.

It is interesting to examine the tradeoffs regarding the state storage location in a sufficiently complex Web services application. Stateful interactions involve inherent URL encoding size limits that are difficult for complex Web services to conform to. The use of HTTP, cookies, and post data for stateful services creates issues in terms of intermediaries, caching, security, and inspection. Another underlying protocol may not have HTTP cookies or post data. Thus, encapsulating state at the transfer protocol becomes a problem. Surfacing this state to the SOAP layer using reference properties helps in solving the problem.

For stateful services, scalability is a major issue. Scaling a stateful service in a server farm environment can be a big challenge. One solution is to continue tunneling client interaction to the original service endpoint inside the service farm. Another is to enable the farm to migrate state or make state available between the different nodes of the farm. The tradeoff analysis between stateful services and stateful interactions in the context of scalability include:

1. **Network costs.** Latency and bandwidth create usability and infrastructure costs in sending the state back to the client. The data plans of some carrier networks may also involve monetary costs for the user of the client device (e.g., mobile phone).
2. **Idling costs.** Cost of dedicated service resources waiting for response in a static dispatching model, or the cost of migrating state between nodes in a dynamic dispatching model.
3. **Caching costs.** For dynamic dispatching service farms, the costs of caching state to a persistent store on the backend, garbage collection for resources, resurrection of resources instantiated with cached state.

Note that for Web services state management, there are also some competing specifications that go much deeper into the above issues. In fact, WS-Addressing will mainly enable technology for these specifications rather than define and describe state management itself.

4.5.1.3 Endpoint Reference Syntax

Now that we have examined the rationale for having EPRs, let us see how they are structured. Essentially, EPRs are quite straightforward. They are comprised of the Web implementation of an address (a single URI) and two optional constructs: `ReferenceParameters` and `Metadata`. We have already discussed the function of reference parameters, but not their structure. The URI address is actually an Internationalized Resource Identifier (IRI). IRIs are the Web's latest notion of an address, a URI update that properly tackles internationalization issues.

Below is the XML Infoset representation of an EPR type taken from the Core specification. As will be shown in the next section, these are abstract types instantiated by message addressing properties such as `ReplyTo` and `FaultTo`.

```
<wsa:EndpointReference>
    <wsa:Address>xs:anyURI</wsa:Address>
    <wsa:ReferenceParameters>xs:any*</wsa:ReferenceParameters> ?
    <wsa:Metadata>xs:any*</wsa:Metadata>?
</wsa:EndpointReference>
```

Example 4-4. An endpoint reference XML Infoset.

An EPR is comprised of the following elements:

- The `Address` is an IRI representing the address of a service endpoint. It identifies the target domain or device and enables the identification of service-level dispatching components. The granularity of a given IRI's addressing semantics is entirely up to the implementation.
- `ReferenceParameters` form an optional construct, which according to the Core specification can be any number of parameters that facilitate proper interaction with the endpoint. That is a rather open definition for performing almost anything. Specifications layered on top of WS-Addressing in the Web services architecture may define specific reference parameters and constrain them to be used in a particular fashion. In addition, reference parameters are namespace-qualified, which allows them to be mapped to specific semantics. Each parameter is unconstrained in the depth of the information model it contains. Reference parameters are created by the EPR developer for interaction with the service and are therefore to be completely opaque to all service users.
- `Metadata` is also unconstrained in terms of structure. Its purpose is to describe the behavior, policies, and capabilities of the endpoint. `Metadata` can be used to ease the processor's burden of separately retrieving an endpoint's policy. It can also be used if the capabilities and constraints of the endpoint are issued dynamically.

The address IRI is unconstrained by the WS-Addressing Core schema document. Indeed, it can be any address allowed by the IRI specification. This means that the layering between device and service semantics is up to the developer of the EPR.

The WS-Addressing Core specification predefines two addresses and their associated semantics. From the perspective of the EPR schema, they are unconstrained and can be used in any message addressing properties instantiation of an EPR. However, they only make sense in particular situations. The following IRI addresses are defined:

1. `<http://www.w3.org/2005/08/addressing/anonymous>` is used when a meaningful IRI cannot be used to locate a service. The WS-Addressing protocol binding (e.g., the SOAP binding of WS-Addressing) is then used to define the precise behavior of this IRI. For instance, a mobile device performing a synchronous request-response message exchange might not have reachable IRI. Thus, it would use the anonymous address, and the SOAP binding would indicate that the underlying SOAP protocol binding must provide a response channel to the resource. The use of the anonymous address enables a SOAP processor aware of WS-Addressing to make use of the underlying protocol's semantics without exposing them at the SOAP layer. This tunneling effect should be limited to cases in which creating a valid, reachable EPR is problematic.
2. `<http://www.w3.org/2005/08/addressing/none>` is used to indicate that the endpoint cannot send or receive messages. This is useful in situations where the service provider creates and distributes an EPR to its service endpoint and wants to positively communicate that neither reply or fault messages will be sent by the endpoint.

Our earlier WS-Addressing example used reference parameters to encapsulate a mobile endpoint for the SMS gateway intermediary. This enabled the gateway to make the protocol change from HTTP to SMS and target the mobile device. The dispatch data included both the SMS identifier of the mobile device and its service node – the SMS port listening for messages. Alternatively, one can think of the SMS gateway as a mobile proxy rather than as an intermediary. In that case, reference parameters can be used as the dispatch data needed to reach the mobile device, as the SMS gateway is no longer a SOAP intermediary but the ultimate receiver.

Each of the described mobile application examples provides a suitable deployment model and an accompanying EPR. However, it is unfortunate that there are no direct semantics available for intermediaries. Interestingly, the precursor specification of WS-Addressing, called WS-Routing, did address SOAP intermediaries and the routing of SOAP messages. However, WS-Addressing and EPRs create an ideal foundation for a new Web services specification designed to route messages through SOAP-layer intermediaries.

How `Metadata` will be fully exploited remains to be seen, but it will certainly be a mechanism to tag a service endpoint with policy constraints and WSDL information. At the time of the standardization of WS-Addressing, the Web Services Policy Framework (WS-Policy) specification had not been submitted to a standardization body. As a result, the WS-

Addressing language must be agnostic regarding WS-Policy, following the W3C practice of not referencing in a normative fashion a body of work which itself is not a normative product of a standards organization. The submission of WS-Addressing did normatively link WS-Policy to EPRs. One future possibility is that once WS-Addressing and WS-Policy have both been standardized, the WS-I organization will profile the use of WS-Policy assertions in the context of WS-Addressing metadata.

4.5.2 Message Addressing Properties

Message addressing properties introduce the syntax and semantics of message exchange to the baseline EPR definitions. This combination results in a base set of addressing information needed for transport-independent SOAP messaging. Message addressing properties are used to define the last leg of dispatch, message correlation, and applications of EPRs relative to some basic, well-known message exchange patterns.

The following XML Infoset representation of the message addressing properties is taken from the Core specification:

```
<wsa:To>xs:anyURI</wsa:To> ?
<wsa:From>wsa:EndpointReferenceType</wsa:From> ?
<wsa:ReplyTo>wsa:EndpointReferenceType</wsa:ReplyTo> ?
<wsa:FaultTo>wsa:EndpointReferenceType</wsa:FaultTo> ?
<wsa:Action>xs:anyURI</wsa:Action>
<wsa:MessageID>xs:anyURI</wsa:MessageID> ?
<wsa:RelatesTo RelationshipType="xs:anyURI"?>xs:anyURI</wsa:RelatesTo> *
<wsa:ReferenceParameters>xs:any*</wsa:ReferenceParameters> ?
```

Example 4-5. The message addressing properties XML Infoset.

Message addressing properties are comprised of the following elements:

- The `To` element indicates the destination of the current message. If not present, the default value is <http://www.w3.org/2005/08/addressing/anonymous>. This element will take on the value of the Address property of the EPR that the message is being sent to.
- The `From` element is an optional property indicating the source EPR which created the message.
- The `ReplyTo` element indicates the reply EPR for the receiver of the current message. If not present, the default EPR's `Address` value is <http://www.w3.org/2005/08/addressing/anonymous>. If the address value is <http://www.w3.org/2005/08/addressing/none>, then no reply message is sent. A reply to this message would use the `ReplyTo` EPR to

create the corresponding message addressing properties. This EPR's `Address` would then be used in the `To` element, and its reference parameters would be copied directly to the corresponding element of the reply message's message addressing properties. The EPR's metadata would be processed to guide the reply message creation.

- The `FaultTo` element indicates the fault endpoint for the receiver of a fault of the current message. The default behavior is to use the `ReplyTo` EPR, if present. The `FaultTo` element is processed like the `ReplyTo` element, except that the former is triggered by a fault condition. The fault message's message addressing properties are mapped like those of the `ReplyTo` element described above.
- The `Action` element contains an IRI identifying the message semantics and target operation. It is used by the receiving SOAP processor for operational dispatching. In addition, the `Action` element can provide a link to a WSDL description of the service.

We have discussed SOAP dispatching and how EPRs facilitate the process. The last stage in SOAP dispatching is getting the service operation identified, located, and invoked. Like the dispatch information modeling offered by EPRs, this last activity has historically depended on the kindness of the service platform to reach into the transfer protocol for operation identification rather than having it being defined explicitly at the SOAP layer. SOAP operation identification at the transport layer has typically been done using the HTTP action parameter in SOAP 1.2 or the `SOAPAction` HTTP header in SOAP 1.1. SOAP applications that use a transfer protocol other than HTTP must find some other means in the transport to communicate the value and semantics of the action feature of SOAP.

Consequently, including the `Action` parameter in the message addressing properties is a main layering feature of WS-Addressing. The parameter enables the fine-tuning of dispatching to the correct service component to be done completely at the SOAP layer. Additionally, the parameter is recommended to be either explicitly or implicitly associated with the service's WSDL definition. The Core specification recommends that the parameter's IRI value is the same IRI identifying an input, output, or fault message in a WSDL interface (WSDL 2.0) or port type (WSDL 1.1). In each case, the parameter is associated with a WSDL operation. The following is an explicit specification of the `Action` parameter in a WSDL 2.0 description:

```
<definitions targetNamespace="http://calendar.nokia.com" ...>
  ...
  <interface name="SynchronousCalendarSchedulerInterface">
    <operation name="NewCalEvent" pattern="http://www.w3.org/2004/08/wsdl/
in-out">
      <input element="tns:NewEventInput" messageLabel="In"
            wsa:Action="http://calendar.nokia.com/new-event-request"/>
      <output element="tns:NewEventOutput" messageLabel="Out"
            wsa:Action="http://calendar.nokia.com/new-event-response"/>
    </operation>
  </interface>
  ...
</definitions>
```

Example 4-6. The WSDL 2.0 definition of the Action parameter.

This WSDL snippet describes a simpler calendar scheduler service than the example presented in the beginning of the chapter. The service described here is of the request-response type, while the earlier example was an asynchronous callback service. Here, the `Action` parameter is explicitly tagged to the associated WSDL operation. With this explicit association, SOAP-layer dispatching of the message to the proper handler of the `NewCalEvent` operation can now be easily supported by SOAP tools and platform runtime.

- The `MessageID` element is an absolute IRI uniquely identifying the message. The sender is responsible for the uniqueness of the IRI. The IRI is used for message correlation.
- The `RelatesTo` element has the cardinality of zero-to-many and is used to provide a log of messages (by `MessageID`) that are related to this specific message. The Core specification defines an abstract property called relationship at the abstract layer. Each `MessageID` in the `RelatesTo` log is associated with particular relationship semantics. One predefined sample of relationship semantics is provided in the Core specification: `<http://www.w3.org/2005/08/addressing/reply>`. This fits the limited scope of the WS-Addressing specification of explicitly handling only one-way or request-response message exchange patterns. A one-way message involves no reply message and thus no relationship semantics. The reply message to a simple request-response message would use the above predefined semantics. They are also used if the `RelationshipType` is not defined. Let us look at two examples:

```
<wsa:MessageID>http://cal-service.com/asdfasdf</wsa:MessageID>
<wsa:RelatesTo>
    http://cal-service.com/12345
</wsa:RelatesTo>
```

```
    <wsa:MessageID>http://cal-service.com/asdfasdf</wsa:MessageID>
    <wsa:RelatesTo RelationshipType=http://www.w3.org/2005/08/addressing/
reply>
        http://cal-service.com/12345
    </wsa:RelatesTo>
```

Example 4-7. Creating a relationship value pair for a one-way message and for a request-response message.

Both code snippets produce the following relationship value pair: `<http://www.w3.org/2005/08/addressing/reply>` and `<http://cal-service.com/12345>`.

The Core specification includes extensibility points that allow another specification author or application developer to add other semantics to this Core specification. One extensibility point involves the `RelatesTo` relationship. The relationship semantics form an attribute named `RelationshipType`, which has an IRI as its value. The following sample extension is for a publish-subscribe service (which is essentially an in-multiple-out message exchange pattern), where we define new relationship semantics to be used in `RelatesTo`:

```
    <wsa:MessageID>http://cal-service.com/qwertyqwe</wsa:MessageID>
    <wsa:RelatesTo>
        RelationshipType=http://example.com/addressing/subscribed
        http://cal-service.com/abcde
    </wsa:RelatesTo>
    <wsa:RelatesTo>
        RelationshipType=http://example.com/addressing/subscribed-event
        http://cal-service.com/98765
    </wsa:RelatesTo>
```

Example 4-8. Defining relationship pairs for a publish-subscribe service.

This produces the following relationship pairs:

- `<http://example.com/addressing/subscribed>` and `<http://cal-service.com/abcde>`
- http://example.com/addressing/subscribed-event and `<http://cal-service.com/98765>`

Thus, the `RelatesTo` feature can enable correlation capabilities for increasingly complex message exchange patterns.

- `ReferenceParameters` are the reference parameters of the EPR being addressed in the message. In the WS-Addressing world, a message is sent by targeting a specific EPR. This may be a `ReplyTo` or a `FaultTo` situation, or a situation in which a service provider creates an EPR to which users can address messages. This EPR is probably associated with the service's WSDL file. For more complex message exchange patterns, a new message addressing property may be created with an EPR as its value. In all of these cases, the reference parameters are identical to each of that EPR's reference parameters. All children, attributes, and in-scope namespaces are represented as they are in the originating EPR. Likewise, as indicated above in the description of the `To` property, the EPR's `Address` property is copied into the `To` field.

4.6 Binding Specifications

WS-Addressing contains two binding specification documents. One is for SOAP and the other for WSDL. Most of the core concepts related to these documents have already been described above. The documents are intended more for Web services platform and tool vendors than for the potential service or application developer. They describe in detail how to make the capabilities described in the Core specification actually work with key Web service technologies.

The SOAP binding specification lays out the Core capabilities in SOAP 1.2 terminology – those of features and modules. A SOAP feature is an abstract representation of a new set of capabilities and each feature set is identified by a URI. For WS-Addressing, the URI is `<http://www.w3.org/2005/08/addressing/feature>`. Each abstract feature has abstract properties. For the WS-Addressing feature, the properties are the same as the message addressing properties: `ReplyTo`, `Action`, etc. Each SOAP message exchange pattern can use this new feature if desired.

Features may collide at an abstract level. If so, a clarifying language is needed. One collision is clarified in the SOAP binding regarding the SOAP Action feature defined in SOAP 1.2.

In SOAP 1.2, a SOAP feature is implemented by a SOAP module. This SOAP Binding specification maps each of the addressing feature properties to a corresponding SOAP header. For `ReferenceParameters`, this SOAP module defines an additional Boolean attribute: `IsReferenceParameter`. This attribute must be used in all reference parameter SOAP headers. The SOAP Binding specification further defines the creation of SOAP headers as the binding of the message addressing properties. Thus, we can change the XML Infoset representations of message addressing properties into SOAP header blocks.

The SOAP Binding specification achieves two other important tasks:

1. It clarifies the use of the anonymous address `<http://www.w3.org/2005/08/addressing/anonymous>` to tunnel into SOAP layer's underlying protocol (defined using SOAP 1.2 Part 2 Adjuncts [9]) to provide a channel to the specified endpoint. In the HTTP binding of SOAP 1.2, an anonymous endpoint in a reply message moves the reply message to the HTTP response channel.
2. It defines the details of fault behavior, fault codes, and predefine some faults likely to occur.

The WS-Addressing WSDL Binding is not stable at the time of writing. Thus, describing it in detail could produce a stale exercise for the reader. However, the main functions of the binding are not likely to change: to connect WSDL metadata in EPRs, and to connect WSDL operations to the `Action` message addressing property. We have already discussed the latter function in some detail.

As for WSDL metadata in an EPR, the WSDL Binding specification describes two ways to accomplish the connection: by adding a reference to WSDL metadata, or by embedding the WSDL metadata explicitly. There are three WSDL metadata elements to choose from:

- `wsaw:InterfaceName`: Describes the message sequence for a particular service. This corresponds to a WSDL 2.0 interface or a WSDL 1.1 port type.
- `wsaw:ServiceName`: Identifies the set of deployed service endpoints. The set of endpoints is represented by a service in WSDL 2.0 and WSDL 1.1.
- `wsaw:ServiceName/@EndpointName`: Identifies one endpoint from a set of service endpoints identified by service name. An endpoint is represented by an endpoint in WSDL 2.0 and a port in WSDL 1.1.

Below is an example from the WSDL Binding specification illustrating the reference option. It uses the `InterfaceName` element to provide the metadata connection:

```
<wsa:EndpointReference
    xmlns:wsa="http://www.w3.org/2005/03/addressing"
    xmlns:fabrikam="http://example.com/fabrikam">
  <wsa:Address>http://example.com/fabrikam/acct</wsa:Address>
  <wsa:Metadata
      xmlns:wsdli="http://www.w3.org/2004/08/wsdl-instance"
      wsdli:wsdlLocation="http://example.com/fabrikam http://example.com/
fabrikam.wsdl">
```

```
    <wsaw:InterfaceName>fabrikam:Inventory</wsaw:InterfaceName>
  </wsa:Metadata>
</wsa:EndpointReference>
```

Example 4-9. An EndpointReference element.

Lastly, an emerging area of importance in the WSDL Binding specification is connecting WSDL message exchange patterns (MEPs) to the message addressing properties. The following WSDL MEPs have been mapped to the message addressing properties. The mapping indicates whether each property is mandatory for the MEP, and provides a general description of its usage. Both WSDL 2.0 and 1.2 MEPs are mapped to message addressing properties.

4.7 Conclusions

The main part of the Core specification, the document defining the primary new concepts of WS-Addressing, is only about a dozen pages long. Thus, this primer on the specification is also very brief. However, the brevity of the specification should not be interpreted as it being unimportant. On the contrary, the specification presents a number of implications that are not articulated in the actual text. The earlier example shows how new types of applications can now be realized in a standardized fashion. With the wide adoption of WS-Addressing as the de-facto Web services architecture, there are many efforts that will leverage its capabilities.

STANDARDIZATION ACTIVITY

WS-Addressing was initially a private specification by Microsoft, IBM, BEA, and SAP, and it was first published in 2004 and went through a revision in 2004. It was submitted to the W3C in 2004 as a member submission.[1] An alternative specification, WS-MessageDelivery by, for example, Oracle and Nokia was first published as a W3C member submission in 2004.[2] Since specifications competing in the same or overlapping technical space delay the achievement of interoperability, the W3C chartered a new working group to create a single standard.[3] The chartering took place in 2004, and at the time of writing, the work continues and is planned to be completed in 2005. The charter stipulates that the WS-Addressing specification is to be the basis for new standardization work and the result would retain the name of the submission: WS-Addressing.

WS-Addressing was a single document submission to the W3C. However, the WS Addressing Working Group has published three separate specifications: WS-Addressing Core[4], WS-Addressing SOAP Binding[5], and WS-Addressing WSDL Binding[6]. The Core specification defines the information model, abstract properties, and their processing behavior that together provide the features of WS-Addressing. The SOAP Binding specification defines a SOAP 1.2 binding of the abstract properties defined in the Core specification. This SOAP binding follows the SOAP 1.2 conventions for extending

SOAP capabilities by using the SOAP 1.2 feature and property mechanisms. To be backward compatible with SOAP 1.1 implementations, the binding specification defines a SOAP 1.1 extension to bind all abstract properties defined in the Core specification. Lastly, the WSDL Binding specification defines how the abstract properties can be described using either WSDL 1.1 or WSDL 1.2. It should be noted that from the application developer's perspective, the Core specification is the most interesting document, because it defines and describes the new addressing capabilities. Usually the capabilities of binding specifications are automatically supported by the Web services development environment, toolkits, and runtime libraries.

4.8 Policy

A policy is a written set of rules. These rules define appropriate behavior, and they can be used to clarify decision making. In organizations, policies are used to govern the way the organization operates, and some policies may be enforced through legal action. For example, the bylaws of an organization may define aspects of the organizational structure and operation. Other organizational policies may define how the organization works with other organizations or manages its finances.

Written policies serve more than one purpose. They can be used to create shared understanding among many participants, within and across organizations. They can also be used as a record of decisions and agreements, and as an aid to decision making by providing clear guidance on issues that have already been analyzed and decided at the appropriate level of management. Members of an organization may be held accountable for not following established policies, and various mechanisms may be used to enforce policies.

Organizational policies may involve the use of computer and networking systems (e.g., rules related to maintaining the confidentiality of information). Additional policies specific to computer and network systems may also need to be established (e.g., requirements on reliability and availability). Computer and network system policies may be written in machine-readable format (e.g., by using XML). These policies may be enforced by functionality embedded in the systems while remaining useful for sharing policy information among the users of those systems. The use of policies becomes especially important in Web services because they extend over organizational units and even different organizations.

To summarize, a policy may serve a number of purposes for Web services, enabling the policy writer to:

1. Clearly articulate rules for human review and understanding
2. Associate rules with appropriate targets, such as resources or actions
3. Enforce rules at an interface, such as a communication endpoint or a service interface
4. Achieve agreement on policy among communicating parties, enabling interoperability
5. Support negotiation on a commonly agreed set of rules, when more than one policy may apply and be acceptable

4.8.1 Policy Types

For Web services, two important policy types are authorization policies (access control) and policies used to reach agreement in messaging and system integration. The latter includes domain-specific policies such as a security policy.

4.8.1.1 Authorization Policy

A service provider must make decisions whether to grant a requester access to a given resource. The access may allow the viewing or modification of data, or the execution of some action. Examples include viewing the geographic location of a mobile user, changing an attribute such as contact information, or initiating an action such as a fund transfer. Making no decision is the same as making a fixed decision (e.g., grant access) regardless of the information available.

Authorization requires that an authorization enforcement point (such as the service provider's Web service) allows or denies an action based on an authorization decision. An authorization decision should be made according to an authorization policy that uses inputs in combination with a set of rules to make a decision. A simple policy, for example, is to only allow access to the members of an organization. This may mean that the policy constrains the decision point to make the following checks:

1. The request includes a credential stating the identity of the requester.
2. The stated identity corresponds to a known entity in the organization.
3. A trusted authority issued the credential.
4. Various security threats have been addressed acceptably.

The authorization decision can be made at the service provider, or elsewhere, at what is generally called a policy decision point, in this case an authorization decision point. The policy decision point uses the request and other available information to make a policy decision, and conveys that decision to the policy enforcement point, which then allows the request to be processed or denies further action. The policy enforcement point is implemented at the service provider.

The authorization decision policy may be used in an automated manner by the policy decision point, but it may also be used for sharing and reviewing the policy among members of the organization. Such policy statements can be fairly complex and may require a language capable of expressing various requirements. A standard related to the authorization process is the OASIS eXtensible Access Control Markup Language standard (XACML).

4.8.1.2 Domain-Specific Policies

A policy may also apply to a specific domain or area of Web services, such as message communications. Such policies outline the specific requirements for a given area, and the sender and/or receiver might not agree to communicate unless their policies in such areas were compatible. These policies include, for example, the following domains or areas:

- Messaging
- Addressing
- Reliable messaging
- Security

To give some examples, policies may require:

- Adhering to the WS-I Basic Profile to enable successful SOAP *messaging*
- Requiring `MessageIDs` (WS-Addressing), to enable explicit correlation of responses with requests and thereby successful *processing*
- Defining requirements for timeout and retransmissions (Web Services Reliable Messaging Policy Assertion) to enable reliable *messaging*
- Detecting inappropriate changes in messages and preventing inappropriate viewing of messages in transit (WS-Security) to ensure *security*

One specification in development for enabling domain-specific policies to be expressed in a generic framework is the Web Services Policy Framework (WS-Policy). A related specification for expressing security policy is the Web Services Security Policy Language (WS-SecurityPolicy). However, it should be noted that these are draft proprietary specifications that have not yet achieved open standardization.

4.8.1.3 Policy Interoperability

A significant source of the value of Web services is interoperability (others include, for example, dynamic composability and extensibility). Interoperability can be achieved to a large degree by agreeing on the use of open standards, and agreeing on the profiles of those standards (where such standard profiles exist). For example, the WS-I Basic Profile is used to profile the SOAP standard. Even so, additional agreements are required on the details of the mechanisms chosen to fulfill policy requirements.

One example is a communication policy that requires a message to be secured so that it cannot be easily modified without detection or have its contents viewed inappropriately while in transit. Conceptually a receiver that does not agree with the communication mechanisms may deny access, in effect constituting an authorization policy that can be enforced at the networking portion of the architecture. Since a variety of security mechanisms can be used to meet security requirements [SecurityChallenges], policies should be able to allow alternative mechanisms to be used as long as they meet the security requirements. This may require policies to be examined for compatibility. The following example outlines ways to express policy requirements for message integrity and confidentiality.

Effective Web services require interoperability at the policy level, which means that participants need to be able to ensure that their policies are compatible for a Web service to function. There are different ways to ensure the compatibility of policies, such as the following:

- Out of band agreement
- Sender assures compatibility with stated policy of Web service (policy attached to WSDL)
- Negotiated agreement in message protocol

With closed communities, such as an enterprise and its partners, it is possible to compare and review policies for compatibility as part of the process of establishing relationships. This does not have to be automated and can be similar to the process used to establish legal and business agreements. In many cases, an organization may even mandate the policy to be observed by its partners.

Another approach is for a Web service provider to document the policy requirements in a machine-readable manner, and associate that policy statement with the Web service description (e.g., the WSDL description). This approach requires the client to obtain the policy as part of obtaining the service description, and ensuring that the client observes the policy when contacting the server. This emerging approach requires standardization of the means of expressing these policies and attaching them to the WSDL description before wider adoption is possible. One approach of this type is outlined in Web Services Policy Attachment (WS-PolicyAttachment), a specification that at the time of writing has not yet been submitted for standardization.

The Liberty Alliance Web Services Framework supports the notion of Web service consumers contacting a discovery service to learn the location of an appropriate Web service provider. This discovery service not only returns an EPR for the service and security tokens needed to access that service, but also mediates the policy requirements of the service. For example, it can include policy requirements in the EPR returned to the Web service consumer. Mediated access can be used to implement policy enforcement.

The most generic approach is for a Web service consumer and provider to negotiate a mutually suitable policy. One method is for the consumer to state policy requirements and capabilities in a Web service request and for the Web service provider to respond with an alternate policy proposal, if it does not accept the policy. Another related approach is for the Web service provider to merge the policies and determine a suitable agreement. An approach using the XACML standard has been proposed. Approaches like these are additional topics for ongoing development of Web service policies.

4.8.1.4 Web Service Policy Mechanisms

One approach to Web service policies is to define a generic policy mechanism that can be used in conjunction with one or more domain-specific policy assertions. In this case, a generic policy language can be used to combine one or more domain-specific policies with generic policy statements, and a generic mechanism can be used to attach this policy to messages and/or Web service descriptions. This is the approach taken with the WS-Policy set of specifications, which include generic assertion statements and mechanisms (WS-Policy), generic means for associating policy with description and discovery services (Web Services Policy Attachment or WS-PolicyAttachment), and domain-specific policy assertions (WS-SecurityPolicy and WS-ReliableMessaging). Note that for such an approach to work, policy assertions must be atomic so that a generic policy mechanism can be used without concern for the details of the individual assertions. Other approaches have also been proposed that effectively allow merging polices to obtain compatible policies (WSPL).

A policy can be associated with the description of a Web service, allowing potential service clients to better understand the requirements of the service, which could mean, for example, including the necessary credentials and materials when making Web service requests to that service.

To give a concrete example, a Web service may require the integrity of the SOAP headers and body of service messages to be protected, and the confidentiality of the SOAP body to be ensured. The service provider's policy may allow SSL or SOAP Message security (or both) to be used for this purpose. When SOAP Message Security is used, the server may require the use of either the OASIS Web Services Security X.509 or SAML token profiles.

WS-Policy provides a framework for expressing policy assertions, where assertions concerning security represent one type of assertion. The following example uses WS-Policy to define integrity and confidentiality assertions.

NOTE

WS-Policy has not been submitted to a standards organization, so the details of this material may change by the time it becomes a standard. Some specifications related to WS-Policy for specific assertions are being submitted to standards bodies, but again, they may change because we are still early in the standardization process. Due to the early stage, some of the specifications may also appear inconsistent. Therefore, the examples in this section only provide an overview, not the exact details.

WS-Policy provides ways to combine policy assertions. Therefore, we use the following pseudo code to express the assertions of a sample policy:

```
<wsp:All>
 Integrity
 Confidentiality
</wsp:All>
```

Example 4-10. A pseudo code snippet illustrating integrity and confidentiality assertions.

In this example, `Integrity` and `Confidentiality` are placeholders for the integrity and confidentiality assertions used to specify that both of these security requirements must be met.

The WS-Policy specification defines a simple way to combine policy assertions, a way to express policy statements in a compact form, and a way to normalize policy statements. These mechanisms are described in the WS-Policy specification [WS-Policy].

Assertions can apply to different areas, such as security or reliable messaging, and are defined in different specifications. Details on assertions related to security are specified in WS-SecurityPolicy.

The `SignedParts` assertion can be used to express the requirement for integrity. Without additional optional information, this assertion specifies that all the headers targeted in the SOAP Ultimate Receiver and in the SOAP body are to be integrity-protected. This assertion does not specify how the integrity requirement is to be met – SSL/TLS or SOAP Message Security (or both) can be used.

Similarly, the requirement for confidentiality can be expressed using the `EncryptedParts` assertion. Without additional information, this assertion specifies that the SOAP body must be confidentiality-protected, but additional headers may be specified as well.

Thus, we could use the following policy statement to specify that the integrity of the body and the headers must be protected and that the confidentiality of the body must also be protected:

```
<wsp:All>
  <sp:SignedParts />
  <sp:EncryptedParts />
</wsp:All>
```

Example 4-11. Illustrating messaging security options.

If SSL/TLS is used with appropriate ciphersuites, and there are no SOAP intermediaries on the message's transport path, then this requirement can be met. The alternative would be to use SOAP message security.

WS-SecurityPolicy also allows assertions to be made regarding the tokens used in conjunction with SOAP Message Security. The following statement could be made to specify the use of either an X.509 or SAML token:

```
<wsp:ExactlyOne>
<sp:X509Token />
 <sp:SamlToken />
</wsp:ExactlyOne>
```

Example 4-12. Choosing a token to use.

Each of these tokens could specify detailed requirements on the properties of the used token.

Thus, the following policy statement could be made to meet the originally stated integrity and confidentiality requirements:

```
001      <wsp:All>
002        <sp:SignedParts />
003        <sp:EncryptedParts />
004        <wsp:All>
005          <wsp:ExactlyOne>
006            <sp:X509Token />
007            <sp:SamlToken />
008          </wsp:ExactlyOne>
009          <sp:httpsToken wsp:optional="true" />
110        </wsp:All>
111      <wsp:All>
```

Example 4-13. A complete policy definition about integrity and confidentiality requirements.

Lines 1, 2, 3, 5–10 state that the integrity, confidentiality, and mechanism requirements must all be met. The `All` element between lines 4–10 defines the mechanism requirement as either the use of an X.509 or a SAML token in conjunction with SOAP message security, or the use of HTTPS (SSL/TLS), or both. If necessary, details on how the tokens are to be used could be defined in the assertions on lines 6 and 7.

The method used to associate a policy statement with a Web service is described in the WS-PolicyAttachment specification [WS-PolicyAttachment].

A policy statement can be associated with, for example, a WS-Addressing endpoint reference or a WSDL definition of a service endpoint.

Association with an EPR can be stated with a `PolicyAttachment` element:

```
<wsp:PolicyAttachment>
 <wsp:AppliesTo>
   <wsa:EndpointReference> ... </wsa:EndpointReference>
 </wsp:AppliesTo>
 <wsp:PolicyReference URI="uri for policy statement" />
</wsp:PolicyAttachment>
```

Example 4-14. Associating a policy with an EPR.

The policy association must be obtained before the association of the policy with the endpoint can be understood.

Another approach is to associate policy statements with the WSDL description of a Web service, for example, with the `portType` definition of the WSDL description:

```
<portType name="SampleOperationPort"
          wsp:PolicyURIs="http://www.example.com/sample-policy" >
    <operation name="SampleOperation">
      <input message="tns:SampleOperation"
             wsp:PolicyURIs="http://www.example.com/sample-policy" />
      <output message="tns:SampleResponse">
    </operation>
</portType>
```

Example 4-15. Associating policy statements with a WSDL `portType`.

More detailed information is provided in the WS-PolicyAttachment specification, including details on the use of WSDL and UDDI.

The OASIS XACML standard is an open standard suitable for creating policies related to authorization decisions. Other mechanisms for defining domain-specific policies, such as security policies, are currently proprietary specifications, which have not been standardized. The same is true of mechanisms for associating policy with WSDL and SOAP messages. An industry agreement on open standards for Web service policies in the near term would further drive the adoption and interoperability of Web services, especially across organizational boundaries.

Chapter 5: Identity and Security

This is the second-to-final chapter in our introduction to Web service technologies, and we are gradually moving toward more sophisticated technologies. Here, we introduce advanced concepts related to identity-based Web services and to techniques involved in establishing identity, as well as security mechanisms used to protect the integrity and confidentiality of Web services. We conclude this chapter by presenting some specific issues related to mobile applications.

5.1 Identity and Web Services

Establishing and conveying identity information is essential to meaningful Web services. This is necessary in order to establish trusted relationships, confirm the source and integrity of information, increase the usability of the system, and enable services that can use identity-based information to offer value-added services. This section discusses how identity can be established through authentication, how identity information can be shared, how permission-based attribute services can protect privacy while making information available, and how services can be enhanced with identity information.

It is often important not to disclose information associated with individuals (*personally identifiable information, PII*). This can be required by legal statute, for example by European Union legislation or by regulations in various industries such as financial services or health care. An important concern is the ability to correlate information from different sources to learn personal information. Although confidentiality is a major component of privacy, it is not the whole story. Given the opportunities for information correlation and cooperation among different parties to learn private information, the entire service infrastructure must be designed to protect privacy.

Mobile Web services raise the stakes regarding private information, since mobile devices often contain private information that can be exposed as services (e.g., contacts, calendar, and to-do lists). In addition, the mobile Web services infrastructure enables the tracking and storing of private information, such as the geographic location or the network presence of a user. The infrastructure also enables consolidated billing for services. This section outlines the important area of identity and services related to identity.

5.1.1 Authentication – Establishing Identity

Establishing identity with people you know is not very hard to do in person. People look at you, hear you, and know you well enough to quickly detect an impostor. It is possible, however, for an impostor to succeed in person if he is unknown to the person checking his identity. This is why we need certificates such as passports, driver's licenses, and employee badges. These certificates generally bind information about a person (such as their name, identity, or entitlements) with information that can be used to uniquely identify the person, such as a picture and a signature – things we assume to be relatively difficult to forge.

On the Internet, as a famous New Yorker magazine cartoon put it, "nobody knows you're a dog." It is easy to pretend to be someone else over a distance with unfamiliar parties and with little real interaction. Authentication means using techniques to establish identity, based on something you know, have, or are. A username and a password are something you know, and if you keep your password secret, it is very difficult for others to impersonate you. A token, such as a *Subscriber Identity Module (SIM card)* in a mobile phone, or a card that generates *one-time passwords* (e.g., an RSA SecurID card), is something you have. If it is hard for others to replicate such a device, having one allows you to establish your identity to others, assuming you do not lose the device or share it with someone else. A *biometric*, such as a fingerprint or retinal eye scan, is something you are. It is difficult for others to copy, unless they are able to copy the dataset produced in reading the biometric (or make a cast of your finger in exactly the right way). To minimize the risks associated with these authentication mechanisms, more than one mechanism can be used at once. For example, using a password with a SecurID card is a two-factor authentication mechanism, which can reduce the risks associated with the use of either mechanism alone.

In the simplest scenario, one party authenticates himself to another party. An investor identifies their identity to the brokerage firm before using his account, and ensures he is actually interacting with the broker. Similarly, a retailer identifies himself to a manufacturer when ordering replacement stock, and ensures that he is in fact interacting with the manufacturer. Often one party can identify the other by something they know regarding the private agreement related to the transaction. This is information only shared by the two parties. However, in order to reduce the risks associated with attacks where one party impersonates another, it is necessary to use explicit bilateral authentication when performing electronic transactions or information exchange of any consequence.

It is important for communicating parties to be certain about whom they are communicating with. Usually this is taken for granted in face-to-face communication, but the situation is quite different in electronic communication. Communicating with an unknown party may involve great risks. For example, if one party is able to impose as someone else (a method known as spoofing), then message integrity and confidentiality are of no use.

It is the responsibility of the relying party (recipient of a message) to ensure that the authentication level is appropriate to the risks involved. The relying party should perform appropriate signature verification and credentials validation. One example is the validation of an X.509 certificate chain (including revocation status).

A variety of techniques can be used to achieve direct authentication to a relying party. These techniques include:

- Peer authentication over a single hop using SSL/TLS in conjunction with X.509 certificates
- Peer authentication over a single hop using HTTP basic authentication (username and password) or digest authentication, presumably over a secure transport (e.g., SSL/TLS)
- End-to-End Web service client authentication with a Username SOAP Message Security token to convey a username and a password or a one-time password value. This token can be bound to the message with a signature together with a key derivation algorithm, and should be encrypted using SOAP Message Security encryption.

- End-to-End Web service client authentication through a SOAP Message Security digital signature binding a key associated with a SOAP Message Security token to the message. This key can be agreed on in advance, or it can be associated with an infrastructure such as PKI or Kerberos. The key can be associated with an identity using the appropriate token (such as an X.509, Kerberos, or SAML token).
- Passing an authentication assertion (e.g., a SAML token) in the message, binding it to the message.
- Authentication of a mobile device to a mobile network infrastructure using SIM authentication, which effectively uses digital signatures associated with the SIM token key. This is done at the network layer, but it allows a network operator to establish identity and share it with an identity provider (IDP), if the IDP is operating within a mobile operator's infrastructure. Since anyone can pick up and use a mobile device that is not locked, the use of an additional authentication technique may be necessary to reduce risk to an appropriate level.

It can be difficult for every service provider to support the variety of authentication mechanisms listed, and doing so would require a large number of service providers to repeat the implementation. This does not scale. An alternative is for a system component to support authentication as a service, and then provide authentication information to other services. In the *Liberty Identity Web Services Framework (ID-WSF)* this service is provided by an IDP, which creates tokens summarizing information about authentication events. If other service providers can trust these assertions, they no longer have to support a variety of authentication mechanisms.

As a Web service client can use more than one technique to authenticate itself to a Web service provider (or in particular to an IDP), there needs to be a way to determine the used authentication technique. One option is to simply try an authentication method and see if it is accepted, but this might not be useful, especially when various policy constraints associated with a technique must be met. Thus, the option may not be acceptable. Other approaches include deciding the issue out of band, or using a general Web services policy mechanism to reach an agreement. The rejection of an unauthenticated request could include, for example, a policy statement regarding the required mechanism and constraints.

The Liberty ID-WSF Authentication Service Specification defines a general, open approach, in which a series of SOAP messages is used to negotiate an acceptable authentication method. These SOAP messages follow the IETF *Simple Authentication and Security Layer (SASL)* framework and use definitions of authentication methods to allow a negotiation. In this scenario, the desired form of authentication is negotiated using SOAP messages containing `<SASLRequest>` and `<SASLResponse>` elements in the SOAP Body. Supported authentication methods are registered in the SASL framework and have associated identifiers. This information is used as part of the SASL negotiation to establish the authentication method.

5.1.2 Authentication Context

Different means of authentication reduce risks to different degrees. Key factors include the strength of the authentication mechanism itself, how keys associated with cryptographic mechanisms are stored, how strongly the authentication method can be associated with an identity, as well as other factors. In fact, this information can be quantified into levels of authentication strength, and different parties can define a policy indicating the required level authentication. This has been included in the Liberty Authentication Context Specification, which has been submitted to the OASIS Security Services (SAML) Technical Committee and has been incorporated into the SAML 2.0 standard.

An authentication context statement can serve a number of purposes. It can be used as part of an authentication method negotiation to state which forms of authentication are acceptable. The SAML 2.0 Authentication Context mechanism simplifies such negotiation, because it allows categories of authentication strength to be defined (and thus allows participants to ignore the details when they want to). Authentication strength takes into account various factors, such as the authentication method, how keys are stored and protected, the due diligence in associating an identity with a key, and other factors relevant to making authentication effective and meaningful. An authentication context statement can also be stored and used as evidence in, for example, the resolution of a dispute related to non-repudiation.

5.1.3 Web Service Authentication Tokens

Once a Web service client has authenticated itself to an IDP, it can obtain a security token to be used together with SOAP Message Security to enable the client to access a Web service. This token can be included in the header of the SOAP message sent to the service and bound to the message using a digital signature. The service can examine the token and base its authorization decision on the presence and contents of the token and the fact that the service trusts the identity service. The system is able to scale by using tokens issued by identity servers, because each service provider is no longer required to explicitly authenticate each client or support the possibly numerous authentication methods that the IDP supports. To simplify server design, clients can be expected to include appropriate tokens in the headers of SOAP messages.

Consider the following example: An IDP issues a SAML token indicating that the holder of the key (who is associated with the token) has been authenticated. In this case, the SAML authority (for example, an IDP) issues and signs the assertion that the Web service client has authenticated itself. This assertion includes a confirmation method indicating how the service provider should confirm the identity associated with the assertion. In this case, it would indicate that the sender (who is claiming the identity presented in the assertion) should sign the message using the key that is specified in the assertion method. The SAML assertion also specifies other information, such as the validity period of the assertion. It is the task of the relying party (in this case the Web service provider) to check the IDP signature on the assertion, the confirmation method, and other information in the assertion to verify its validity before offering the service.

5.1.4 Replay Attacks

It is important to protect systems from so-called replay attacks. In a replay attack, someone intercepts a SOAP message, makes a copy of it, and then later re-sends the message. Although it might not affect the first message, a replay attack could have serious effects at the application level. For example, if a message requesting the shipment of goods were re-sent to a third party, the result could be two shipments. If portions of the message are not integrity-protected, it might be possible to redirect the additional shipment to a third party, even if the message contained a valid authentication token and signature. The risk is related to the overall use and application of SOAP messages. The problem is that a relying party might not be able to detect the replay attack. If the original message contains a valid authentication token, so can the replay, assuming that the validity period of the authentication token has not expired.

One approach to protecting oneself from replay attacks is to use short validity periods for tokens, and to ensure that the relying party checks that tokens are still valid. This reduces the risk of replay, but does not eliminate it, since there is always a period when a replay attack is possible.

A more powerful approach is to ensure that each message contains unique information, which the receiver can check to detect a replay attack. The information can be part of the message payload, managed at the application level. One example would be a unique transaction

ID, as long as this value is integrity-protected. Alternatively, the sender could use mechanisms at the SOAP message layer. The SOAP Message Security standard discusses this issue, and mentions the following four approaches (chapter 13):

- Timestamp
- Sequence Number
- Expirations
- Message Correlation

To be useful, these approaches may require integrity protection. SOAP Message Security defines a `Timestamp` element that can be used to convey a timestamp as part of a general `<wsse:Security>` SOAP header (which can also be used to convey an authentication token). Thus, the replay risk can be addressed with the appropriate use of the WS-Security `Timestamp` element. (The defined period can be much shorter than the authentication assertion validity period, allowing an authentication assertion to be re-used, but preventing messages that rely upon that assertion).

5.1.5 Single Sign-On

Many organizations want to administer and manage identity information within their own organization. For example, most corporations want to manage the credentials of their employees, taking into account corporate policies and technical requirements. Even if an organization outsourced this function, it would still want to establish the policies and procedures. This means that special consideration must be taken when the service of one organization accesses the service of another, crossing the organizational boundary. The organizations must establish a relationship with each other and create a so-called *circle of trust*, where services can verify the validity of different forms of identification of the other organization, and still effectively authenticate and authorize the other party.

In many cases, a user needs to interact with several Web sites to achieve a goal. For example, to complete a travel reservation, the user may have to visit the Web site of an airline, a hotel, and a car rental agency. If each site requires authentication, accessing them can be annoying and cumbersome. In addition, this may contribute to a security risk, if the user chooses the same password for each account. Repeated authentication can be a security problem for a mobile device user, because the device's user interface can be limited (e.g., small keypad and display). As a result, it is difficult to enter a password that is long enough and difficult enough to be guessed easily. This can be particularly difficult if the user has to repeatedly press a key to cycle through characters. If the user makes the process easier by selecting the first choice of each key, the password can be guessed more easily. Therefore, it is useful to establish identity once with an identity server, and then allow that server to create an identity assertion that can be conveyed to different Web services during a specified period. It is often essential that the users of Web services can mutually verify each other's identity, because this allows them to make the appropriate authorization decisions.

One way to implement an identity federation is to link accounts and establish a name identifier that can be shared across organizations. This name identifier can be linked to a specific local account, which could have a local identifier. The use of different name identifiers for the same entity across different organizational boundaries makes it more difficult for a third party to correlate transactions across these boundaries. The reason for this is that the name identifiers only have meaning to the two organizations (which makes it difficult for a third party to correlate them). Another approach is to use a pseudonym server to assign name identifiers for parties participating in cross-organizational transactions. This method does not require the linking of accounts, but does require the IDP to participate in mapping pseudonyms to verified identities.

In single sign-on (SSO), a Web browser client authenticates itself to an IDP, which then produces an authentication assertion accepted by cooperating service provider sites. This allows the client to access those sites without having to authenticate itself again. A variety of mechanisms have been defined for SSO in the SAML 2.0 specification, converging the earlier standards of the *Liberty Identity Federation Framework (ID-FF)* and earlier SAML standards. SSO requires Web sites to trust and cooperate with an IDP (thereby forming a circle of trust).

5.1.6 Single Sign-On Enabled Web Services

Web services can be integrated with traditional Web browser access. For example, a Web services-enabled client can authenticate itself to an IDP using a browser-based SSO mechanism, obtaining a security token that can be used in SOAP Web service requests. The client may also need to access services on the Web that are not provided using Web services, and it may need to authenticate itself in this process. In this case, the Web service client can authenticate itself first (via the Liberty ID-WSF Authentication Service) to an IDP. After that, the client can contact an ID-WSF Single-Sign-On Service (a Web services interface to the IDP) for a token to be used in browser-based SSO sessions to a Web site.

Another example is a person using a Web browser to access a Web portal. Certain actions on that portal can trigger different types of Web service requests. In this case, an SSO to an IDP via the portal may enable the portal to initiate Web service transactions on behalf of the user (with a security token obtained from the IDP).

5.1.7 Authorization and Audit

Once the identity of a party has been established, it is important to determine what information the party is allowed to access and which actions the party is allowed to perform. This process is known as authorization. It is discussed in the sections on identity services and Web service policy. Keeping action logs is also important for uncovering security attacks as well as for providing a basis for dispute resolution. Records such as logs can be used in conjunction with digital signatures to form the basis for dispute resolution (also referred to as non-repudiation).

5.1.8 Permission-Based Attribute Identity Services

There are two kinds of mobile identity-based Web services. The first is a service that offers identity-related information derived from the mobile infrastructure. The second is a service that relies on another service running on the mobile device itself, offering access to personal information stored on the device.

The identity information service could be offered by a mobile infrastructure operator (for example, to provide information about the mobile device user's physical location). The location information could be used to provide enhanced services, such as weather forecasts, restaurant listings, and other information customized specifically for the user's location. The actual enhancement could be completely transparent to the user.

Consider, for example, a weather information request. In this scenario, the user first sends a request for weather information. This triggers a Web service request for the user's location. The result is that information relevant to the user is utilized to customize a service.

Mobile device-based identity services provide access to personal attributes stored on the device. Modern mobile devices can contain various types of personal information, such as contact and calendar information. In practice, these services could be used, for example, to request free time slots from the person's calendar for booking a meeting. Again, such access requires permission from the user and the authentication of the requester.

Figure 5-1 below shows an example where access to a Web site (Mybirthday.example. com) initiates a Web service request for personal information from the user's personal profile Web service located at MyProfile.com. In this example, Mybirthday.example.com (the service provider) offers a horoscope service, but requires personal information about the user (birthday) in order to provide the service. SSO can be used to obtain access to Mybirthday.example.com, and the site can then use an SSO Web service at the IDP to obtain the authentication token required to access the personal profile Web site.

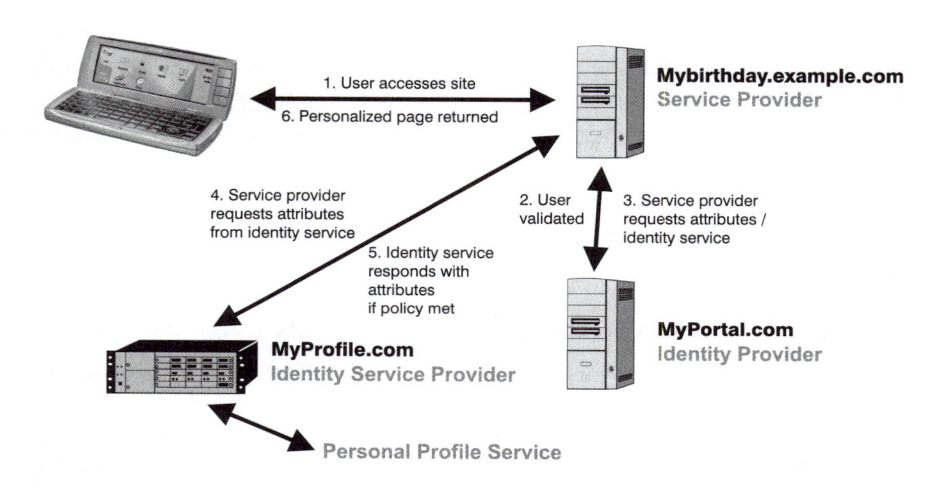

Figure 5-1. Integration of FF with WSF. Source: Rouault, Liberty Alliance: What is Phase 2?

The Liberty Alliance has produced an entire set of standards to address the challenges pertaining to permission-based attribute identity Web services. This set of specifications, the Liberty Identity Web Services Framework (ID-WSF), outlines a system design in which the protection of privacy is an integral part of the architecture, and not an afterthought.

5.1.9 Services Enhanced with Identity Information

Many services can use identity information to provide personalized services. One example is a service that displays time information and automatically adjusts the time zone to match the user's current location.

The Liberty Alliance has developed specifications for creating a Web service invocation framework that supports the use of identity services while effectively protecting privacy at all levels. The SOA for S60 platform and the Nokia Web Services Framework (NWSF) described in chapter 7 support the Liberty Framework as one mechanism to allow applications to invoke various Web services, including those that consume or provide identity information, in a manner that protects privacy and is also easy to use.

5.1.10 Identity Service Description

Although WSDL can be used to describe a Web service, additional information is often required to fully describe a service endpoint. Identification of the necessary security mechanisms and information on how to obtain additional metadata about the service are good examples of additional information useful to a service description.

The Liberty Discovery Service returns the description of an identity-based Web service. The service instance description includes the following description and metadata components:

- **Service Type:** the type of service, identified by a URI that can reference an abstract WSDL document or a Liberty-defined type description
- **Provider ID:** the service provider's URL (location of the service), from where metadata can be retrieved
- **Service instance description:** protocol endpoint and other information needed to access the service, such as that provided by a concrete WSDL document
- **Security mechanism identification:** a URI that indicates the type of security applied
- Security credentials needed to invoke the service

Note that the Liberty Discovery Service does more than just provide reference information on how to access the service: it also authorizes and returns information needed to invoke the service, such as security credentials.

5.1.11 Identity-Based Discovery Services

An identity-based discovery service can be used in identity-based services that require privacy. In this case, a client must authenticate itself to an IDP. As part of that authentication process, it receives information on how to reach its discovery service (a discovery service appropriate to that specific client). The client can then access the discovery service in order to locate a service, but it will only discover services it is authorized to access (services appropriate to the client). This eliminates the problem of a client searching all services and discovering inappropriate ones, and then attempting to access them (an issue that goes beyond efficiency, since it may not be appropriate for the client to even know about the existence of some of these services). Once the client discovers a service in its discovery service, the discovery service returns information necessary for the client to access the service, including security tokens and WSDL description information.

This mechanism is a core component of the Liberty ID-WSF Discovery Service Specification. Like non-identity-based discovery mechanisms, this mechanism returns WSDL information (or a subset of the details expressed in WSDL) related to the services. However, this discovery service only returns service descriptions that the requester is permitted to access, based on the requester's authenticated identity. It also returns the security tokens needed to invoke the service. What is returned is everything needed to invoke one or more identity services (in ID-WSF 1.1 this is called a *resource offering*), which correspond to the original request and which map services to specific resources (or identities). In other words, a description for "Joe's profile service" is returned, rather than a description for a generic profile service, and the description is only returned when it is authorized and appropriate for the requester.

5.1.12 PAOS

The Liberty Reverse HTTP Binding for SOAP Specification (also referred to as PAOS) allows the response to an HTTP request to include a SOAP request. This allows a device that cannot be addressed directly to participate in a SOAP message exchange: the device can initiate an HTTP request, and the recipient of that request can initiate a SOAP message exchange. This is a valuable feature because it allows devices to offer Web services, even if they cannot be addressed directly.

The following example taken from the PAOS specification illustrates the interaction between a device that is browsing a horoscope Web page, and the horoscope Web site itself. The horoscope site can act as a Web service consumer for a Web service running on the mobile device (for example, to retrieve the user's birthday from the device to customize the horoscope).

1. User agent requests a page...

```
GET /index HTTP/1.1
Host: horoscope.example.com
Accept: text/html; application/vnd.paos+xml
PAOS: ver="urn:liberty:paos:2003-08"; "urn:liberty:id-sis-pp:2003-
08", "urn:liberty:id-sis-pp:demographics"
```

2. Server responds by asking for a date of birth...

```
HTTP 200
Content-Type: application/vnd.paos+xml
Content-Length: 1234

<soap:Envelope xmlns:soap="http://schemas.xmlsoap.org/soap/
envelope/">
<soap:Header>
<paos:Request xmlns:paos="urn:liberty:paos:2003-08"
responseConsumerURL="/soap"
service="urn:liberty:id-sis-pp:2003-08"
mustUnderstand="1"
actor= "http://schemas.xmlsoap.org/soap/actor/next" />
<sb:Correlation xmlns:sb="urn:liberty:sb:2003-08"
messageID="6c3a4f8b9c2d"/>
</soap:Header>
<soap:Body>
<pp:Query xmlns:pp="urn:liberty:id-sis-pp:2003-08">
<QueryItem>
<Select>/pp:PP/pp:Demographics/pp:Birthday</Select>
</QueryItem>
</pp:Query>
</soap:Body>
</soap:Envelope>
```

3. Service at user agent responds to the SOAP request with a SOAP response inside an HTTP request...

```
POST /soap HTTP/1.1
Host: horoscope.example.com
Accept: text/html; application/vnd.paos+xml
PAOS: ver="urn:liberty:paos:2003-08"; "urn:liberty:id-sis-pp:2003-
08", "urn:liberty:id-sis-pp:demographics"
Content-Type: application/vnd.paos+xml
Content-Length: 2345
<soap:Envelope xmlns:soap="http://schemas.xmlsoap.org/soap/
envelope/">
<soap:Header>
<sb:Correlation xmlns:sb="urn:liberty:sb:2003-08"
messageID="4d3eae2e3f5g"
refToMessageID="6c3a4f8b9c2d"/>
</soap:Header>
<soap:Body>
<pp:QueryResponse xmlns:pp="urn:liberty:id-sis-pp:2003-08">
<Data>
<Birthday>--05-09</Birthday>
</Data>
</pp:QueryResponse>
</soap:Body>
</soap:Envelope>
```

4. The server responds with a page containing a personalized horoscope...

```
HTTP 200
Content-Type: text/html
Content-Length: 1234
<html>
<head>
<title>Your Horoscope from horoscope.example.com</title>
</head>
<body>
<p>Dear Virgo, <br/>
```

```
In July 2003 you will have to sit through many boring
meetings.
But this ordeal will be worth it and you will make new
friends.</p>
</body>
</html>
```

Example 5-1. PAOS example.

5.2 Security and Web Services

Web service messages can be sent over the Internet and across organizational and geographic boundaries. Therefore, an organization that uses Web services to build an application may have SOAP messages transmitted through an insecure network and through network nodes that are not under its control. It is possible for a node to be hacked in such a way that it can trace, insert, delete, or modify messages that pass through the node, creating a man-in-the-middle attack. This can happen at a protocol layer below SOAP, or at the SOAP layer itself.

Different vulnerabilities can be exploited by different adversaries depending on their capabilities. This section categorizes protection methods that can be used to mitigate the risks associated with different attack levels. However, it should be noted that any party with sufficient funding, experience, and equipment will eventually overcome the protection methods. In addition, the probability of deciphering an encryption increases over time. The techniques described here are intended to protect transactions from common threats over a relatively short period of time. There are many aspects related to the security of important business information, including the following:

- **Audit** – creating records to support dispute resolution (including issues related to non-repudiation)
- **Authentication** – identifying the communicating party
- **Authorization** – determining the entitlements of the party
- **Confidentiality** – taking measures (such as encryption) to ensure that the information conveyed via service messaging cannot be accessed by inappropriate parties
- **Denial-of-service attack mitigation** – taking measures to reduce the impact of attacks designed to disable or reduce the availability of a service

- **Integrity** – taking measures to enable a recipient to detect malicious or accidental changes to a message, including insertion or deletion of messages
- **Replay protection** – taking measures to enable the detection of message re-use
- **Training** – taking measures to avoid social engineering attacks

The previous sections have discussed the important issues of authentication, authorization, and identity management. The following sections discuss issues related to integrity and confidentiality. However, we begin with a brief introduction of the relationship of protocol layers and message flow to these topics. This is followed by a reminder of the importance of mitigating denial-of-service attacks.

5.2.1 Security Services and Protocol Layer Interactions

Security threats can be addressed in different ways at different protocol layers, as shown in Figure 5-2 below.

Layer	Target	Techniques
Application	XML Payload, attachment contents	XML Signature, XML Encryption, CMS
SOAP Messaging	SOAP header blocks, SOAP body, SOAP attachments	WS-Security
Transport	Entire SOAP message and SOAP binding	SSL/TLS, VPN
Link	Entire communication	Dedicated Line

Figure 5-2. Security by protocol stack layer.

Although security techniques can be based on similar mechanisms at different layers, the mechanisms are used differently. For example, the SOAP Message Security layer defines how the XML Signature and XML Encryption technologies are to be used in the context of securing a SOAP message, limiting itself to security processing at the SOAP message layer. An application can apply the same XML Signature and XML Encryption technologies directly on payload content, without the constraints imposed by WS-Security.

Likewise, most security techniques make use of different cryptography algorithms, such as SHA-1 digests and symmetric and asymmetric signature and encryption algorithms. However, the use of these algorithms can differ depending on the protocol layer.

Security techniques can be used simultaneously at different layers. However, this may have an adverse effect on performance and may not be beneficial if adequate security could be obtained at a single layer. In some cases, different security services can be obtained at different layers. For example, SSL/TLS can be used to ensure message integrity and confidentiality for single-hop communication, while authentication can be used at the SOAP messaging layer (e.g., a username and password) by using a username token defined in the OASIS Web Services: Security Username Token Profile.

The appropriate techniques depend on the required level of security. For example, while SSL/TLS may be suitable for protecting communication over a single communications link, this is not the case if a SOAP message travels over more than one link. In that case, the message will be at risk while at the intermediary node, even if SSL/TLS is used for each link. The reason is that SSL/TLS security is terminated at the intermediary node network interface, providing no security at the node itself. This problem is illustrated in Figure 5-3 below.

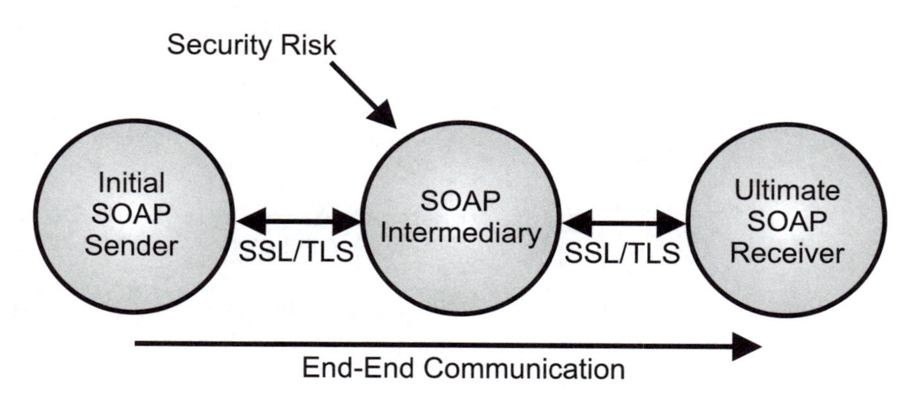

Figure 5-3. Intermediary risk using single-hop transport security.

In some cases, technologies at different protocol layers can be used simultaneously. For example, a mutual authentication technology (e.g., SSL/TLS) could ensure the identities of the communicating nodes, while authentication at the SOAP Message Security layer could use digital signatures to bind the message to end identities. Using both techniques reduces the risk of an inappropriate SOAP node sending a message that is incorrectly interpreted as coming from a trusted node, or a message being sent from an unexpected end entity being incorrectly accepted because it is relayed through a trusted intermediary SOAP node.

It should be apparent that security requires a holistic approach – it is not enough to toss technical solutions together and hope for adequate coverage. Nevertheless, we begin our review from the component technologies.

5.2.2 Message Integrity

Corrupted information can have serious consequences. For example, goods might be shipped to an incorrect address (and subsequently stolen), a financial transaction may be altered (causing financial loss), or operations may be disrupted. Risks posed by these sorts of attacks can be mitigated with the use of cryptographic techniques. These techniques can make it almost impossible to change or substitute information unnoticed. Integrity protection is essential to many Web service applications.

Cryptographic message integrity protects a message from being changed maliciously or accidentally without detection by the relying party. This is different from traditional checksums, which can be used to detect changes, but which can also be replaced by a third party without being detected. Cryptographic message integrity binds a representation of the unchanged message with private information. For this reason, the integrity check cannot be replaced without detection. For example, a symmetric keyed hash (HMAC) is a shortened representation of the message (e.g., a digest) combined with secret key information. This algorithm ensures that the message cannot be changed without changing the HMAC value. It is almost impossible to find a message that corresponds to an arbitrary HMAC. Similarly, a public-private key pair can be used to sign a message digest with similar results.

This technique of creating a short representation of content and either signing it cryptographically or using a secret key is used as a general security technique for message integrity. The sender creates a signature of the material to be protected, and it is the receiver's responsibility to create the same representation and see if they match. If they do not match, the receiver should suspect a breach in message integrity.

The above technique has been incorporated into XML-Signature Syntax and Processing (a W3C Recommendation). The Recommendation outlines ways to protect XML and other content by using digital signatures, and ways to represent the information by using a standard XML structure. This structure includes references to what has been included in the signature, a description of used algorithms, and possibly references to the keys associated with the signature (for example, the public key certificate corresponding to the private key used to sign the message).

The XML signature design is useful for the following reasons:

- It supports a variety of algorithms that are explicitly represented.
- It is designed to work well with XML (for example, the signature result can be incorporated into an XML document, since the signature package is represented in XML).

An application can use XML signatures directly on the payload of a SOAP message, before the payload is processed by the SOAP messaging layer. In this case, the process is transparent to the SOAP messaging layer and must be managed at the application layer. This offers the advantages of the full generality of XML signatures and the ability to integrate such signature processing with other aspects of application processing (including signature storage). Creating an XML digital signature produces an XML block containing information about signed content and used methods, as well as the digest(s) of the content and the signature (see Example 5-2).

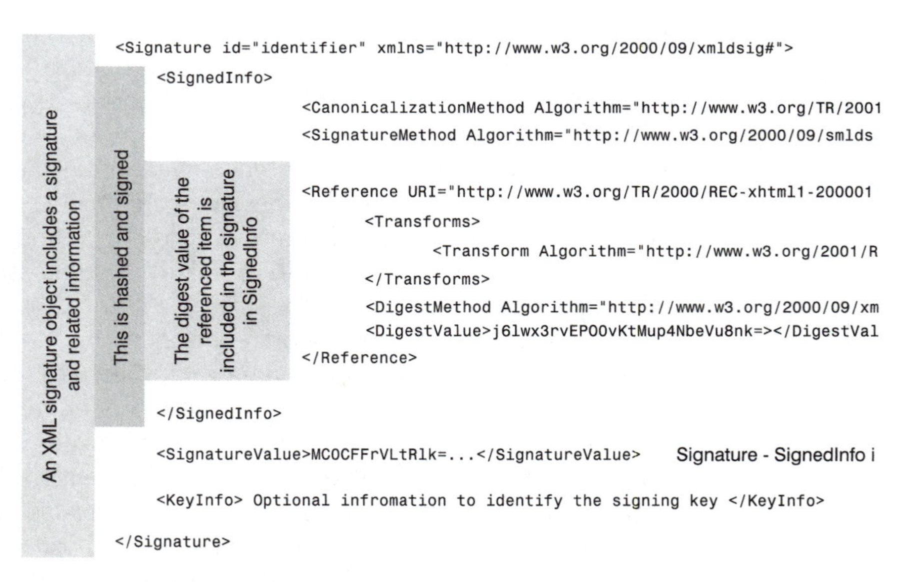

Example 5-2. XML signature structure.

Another approach is to perform signature operations at the SOAP messaging layer, integrating this security processing with SOAP messaging processing. This approach has the advantage that it can be made transparent to the application and integrated with messaging. The signature is not processed directly by using XML signatures, but rather by using WS-Security, which defines how to use XML signatures in the context of SOAP messaging. This approach limits the way XML signatures are used to ensure that they function well with SOAP messaging. For example, signatures and related information are carried in security header blocks with ordering rules designed to ensure that the receiver can process the security components by using previously seen information. The WS-Security standard also defines security tokens that can be used to convey key and other information in an interoperable manner.

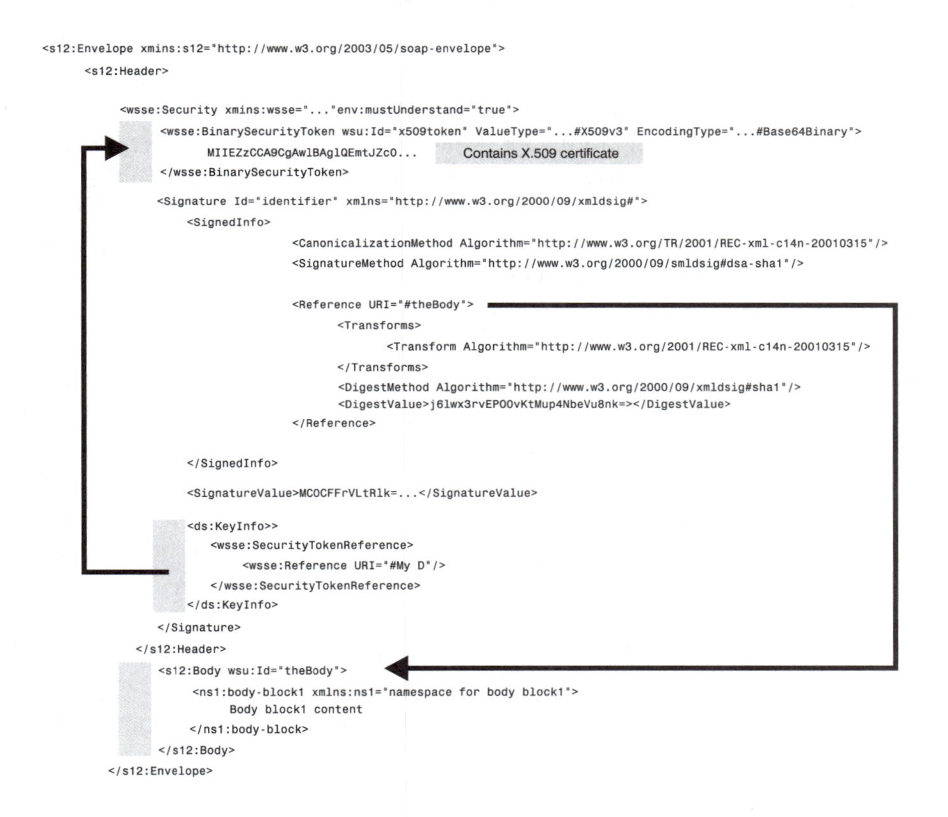

Example 5-3. WS-Security signature structure with an X.509 token.

Integrity protection may be useless if the confidential information used to cryptographically protect the message information is not tied to a trustworthy identity. For example, if the receiver of a signed message does not check that the private key used to create the signature is associated with a trusted partner, then the signature has no security value. A third party could have created the content and then signed it for secure appearance. Thus, integrity is linked to the concept of authentication discussed earlier. Integrity protection is always based on the associated authentication. In addition, if a message can be sent again by a third party (a replay attack), then the message as a whole might no longer be legitimate even though the signature is valid. Therefore, in addition to using digital signatures, it is important to ensure that the signature corresponds to a trusted entity and that the message is meaningful at the time it is received. There is no silver bullet for security problems.

5.2.3 Message Confidentiality

If a message is routed through a node that has been compromised, a third party might gain access to the information (which could be, for example, order, inventory, business plan, marketing, or other sensitive information). The protection of confidentiality is often crucial, depending on the risk associated with the disclosure of the information. Encryption offers a way to significantly reduce the risk of loss of confidentiality.

Encryption can be performed using various shared key or public-private key algorithms. The W3C XML Encryption Recommendation specifies how to encrypt content and package encrypted content using XML. This method allows XML content to be encrypted and replaced with an `EncryptedData` element. It also allows XML with encrypted content to be processed as XML, as long as schema validation is not required (or the schema is very flexible).

XML Encryption enables the encryption of an XML element pair, including the content, or the content without the surrounding start and close elements. The `EncryptedData` element is used to replace the encrypted content, and this structure includes the cipher text (the result of using an algorithm to encrypt the content) as well as information about the encryption algorithm that was used, and possibly information about the key. One common method, called super-encryption, is to use a temporary symmetric key in encryption, and then encrypt this key with an asymmetric key. This provides higher performance while leveraging public key infrastructures for key sharing. In this case, an `EncryptedKey` element can be used to convey the key.

```
<?xml version='1.0'?>
<PaymentInfo xmlns='http://example.org/paymentv2'>
<Name>John Smith</Name>
<EncryptedData Type='http://www.w3.org/2001/04/xmlenc#Element'
xmlns='http://www.w3.org/2001/04/xmlenc#'>
<CipherData>
<CipherValue>A23B45C56</CipherValue>
</CipherData>
</EncryptedData>
</PaymentInfo>
```

Example 5-4. EncryptedData example (from W3C XML Encryption Recommendation).

5.2.4 Mitigating Denial-of-Service Attacks

A service is not useful if it cannot be accessed. A denial-of-service attack is designed to reduce the availability of a service by using service resources (to degrade the service), or by preventing access to the service altogether. The effects of a denial-of-service attack are not easy to mitigate, because it can be hard to distinguish legitimate users from attackers. Some general techniques can be used to reduce denial-of-service threats, such as demanding the requester to do more work in making a request than the receiver needs to in responding to the request (e.g., client puzzles). The authentication of a requester can prevent serious work from being spent on an attacker. A service-oriented framework should take measures to reduce the risk of successful denial-of-service attacks.

5.2.5 Nature of Mobile Web Services

When we consider mobile Web services, certain concepts, particularly those related to identity, become very important. SSO and identity-based discovery increase the usability of mobile devices by reducing the need for explicit authentication. Mobile devices are also becoming especially suitable for acting as providers of identity services, for example by offering personal profile information that is based on information stored and managed on the device (contacts and calendar information).

The following aspects should be considered when designing mobile Web services:

- Identity services are important for mobile Web services.
- Services, which involve mobility and are offered by the mobile infrastructure, can enhance other service provider offerings, by customizing Web services based on location or other information.
- A mobile device can expose personal information via a Web service interface, although measures should be taken to protect privacy. An identity-based Web services infrastructure, such as the one standardized by the Liberty Alliance, is important in this regard, because the protection of privacy requires a specific architecture. The reason for this is that if privacy is lost at any point, other measures will not make a difference.
- A mobile device can offer an interactive service, which is invoked when another service seeks to interact with the user.

The mobile and fixed Internet are converging, and therefore it is natural that core Web services standards are being developed by standards organizations that are not limited to the mobile Internet. These organizations, which include the W3C, OASIS, and WS-I, are discussed in Appendix A. Mobile infrastructure services are being standardized by the Open Mobile Alliance (OMA). This organization has also profiled the core Web services standards produced by other organizations as part of its mobile Web services work.

5.3 Summary

This chapter has discussed key concepts related to identity management and security in Web services, and their implications for mobile Web services. The next chapter takes a more detailed look at a specific identity service – enabling technology portfolio created by the Liberty Alliance.

Chapter 6: Liberty Alliance Identity Technologies

As stated in the beginning of the Identity and Security chapter, the use of identity information is critical to meaningful Web services. The significance as well as the complexity of a user's network identity increases in proportion to the number of services used. This also applies to, for example, the security requirements associated with the services, and features that enable a personalized user experience.

The management of the increasing number and complexity of Web services does not necessarily require a holistic approach on identity management. The problem is that the introduction of new identity and security solutions tied to a specific service provider (SP) and technology creates a fragmented and disparate identity technology landscape, which has negative effects on user experience and security.

To solve this problem, the Liberty Alliance offers a comprehensive standards portfolio that enables identity federation and the providing and consuming of identity-enabled services.[1] These features can be realized by using either traditional HTTP communications and Web browsers, or Web service technologies, and servers and clients based on Web services.

At a high level, the complete specification package consists of three major elements (Figure 6-1):

- **Liberty Identity Federation Framework (ID-FF)** defines a complete architecture for identity federation, single sign-on, and single logout.
- **Liberty Identity Web Services Framework (ID-WSF)** defines a system that enables identity Web services.
- **Liberty Identity Services Interface Specifications (ID-SIS)** define several identity Web services for specific use scenarios. For example, the Liberty ID-SIS Personal Profile Service Specification can be used to create services that manage personal information.

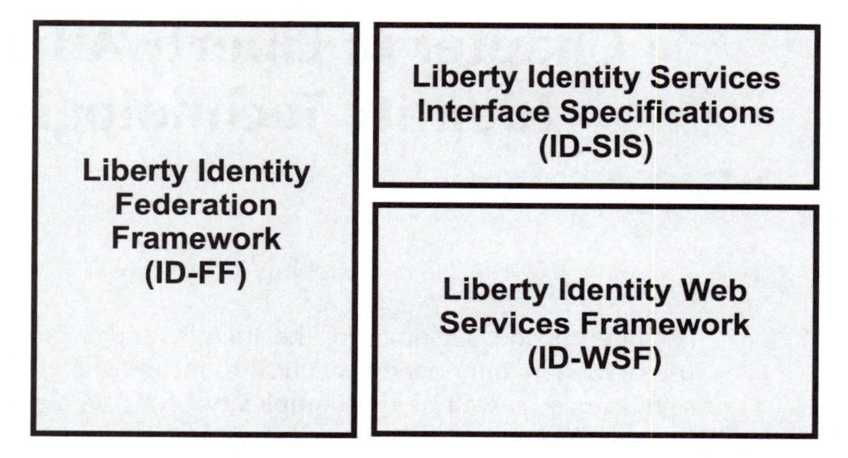

Figure 6-1. The three main sections of the Liberty specification portfolio. Please see the source figure available at http://www.projectliberty.org/resources/specdiagram.php for a detailed view about the relationships of different specifications in the three main areas.

NOTE

For more detailed information on the history and achievements of the Liberty Alliance, see section 4 on Liberty Alliance in Appendix A. For technical details and implementation guidelines, see the Liberty Alliance specifications. This section focuses on the architecture and features that the specifications enable. It should also be noted that because we try to provide the latest possible information on the features of the architecture, some parts of the text are based on draft-level material.

This chapter provides an introduction to the Liberty architecture. It briefly describes the key specifications of the architecture, with the aim of presenting an overview of features instead of bit-level functionalities. To introduce some of the key concepts, and to provide a framework for the discussion on particular specifications, we present two examples scenarios: one for the Liberty Identity Federation Framework (ID-FF) and another for the Liberty Identity Web Services Framework (ID-WSF).

We begin by presenting an example scenario with key players modeled according to the basic features of ID-FF. This example illustrates how a federated identity architecture functions when a user books a flight and a hotel room, and rents a car. The user client is an ordinary Web browser, and the client has user accounts with all of the three SPs.

NOTE

In this book, mobile Web services are described using Nokia mobile devices, and in this chapter, the active, service-consuming party is referred to as the user. In the Liberty specifications, the used term is Principal. The latter term is closer to the truth, because the whole architecture is built around the notion of a principal and the related identity information, without making any assumptions about the realization of this principal, which may be, for example, a human or a service.

The example includes a user, an identity provider (IDP), and three SPs: an airline, a hotel, and a car rental company. The IDP has business and operational agreements in place with all of the SPs, creating a *circle of trust*. From the user's point of view, entities belonging to a circle of trust form a unified system, where all resources are automatically within the user's grasp after one login.

In the first stage, the user creates an account with the IDP. The IDP stores and manages user account data, and implements the actual identity federation features. The account can be protected with a simple username-password combination, or with strong authentication, such as RSA SecurID.

For the sake of simplicity, we assume that immediately after creating the account, the IDP provides the user with a list of SPs within the IDP's circle of trust, and asks if the user wishes to link – or *federate* – the account with the IDP with one or more accounts with SPs. The user selects the desired SP accounts for federation with the IDP, and to implement the actual linking, logs once into each SP account. (Note that the actual implementation may vary.) In the background, the actual federation is realized using the Single Sign-On and Federation Protocol (see section 6.1.1.1).

In the second stage, we proceed to the actual flight reservation in order to describe the single sign-on (SSO) feature. Once the user has logged into the IDP and has proceeded to the airline's ticket reservation service, the reservation service contacts the IDP and requests the user's authentication assertion. From the user's perspective, access to the system is granted automatically. However, in reality the SP uses the Single Sign-On and Federation Protocol to acquire the assertion it needs. The same seamless login procedure takes place with all SPs belonging to the same circle of trust.

In the third stage, the user has completed the reservation and wants to end the browsing session. To expedite the logout procedure, the client uses the single logout feature provided by the IDP. When the IDP receives the logout request, it forwards the request to all SPs for which it has authenticated the user during that session. This is done using the Single Logout Protocol (see section 6.1.1.2).

If the user decides to end the customer relationship with any of the SPs, she can cancel the identity federation for that SP by requesting this from the IDP. The termination is carried out using the Federation Termination Notification Protocol (see section 6.1.1.3).

This basic example scenario did not involve two ID-FF protocols: the Name Identifier Mapping protocol and the Name Registration Protocol. These protocols handle specific technical issues related to federation and identifiers. The protocols are described briefly in the end of section 6.1.1.

One issue worth noting but not covered here is that the Liberty identity federation system is able to handle anonymous access and the use of pseudonyms. These techniques are used to protect the real identity of a user, and still provide access to services that require user authentication.

CASE STUDY: ADDRESS BOOK

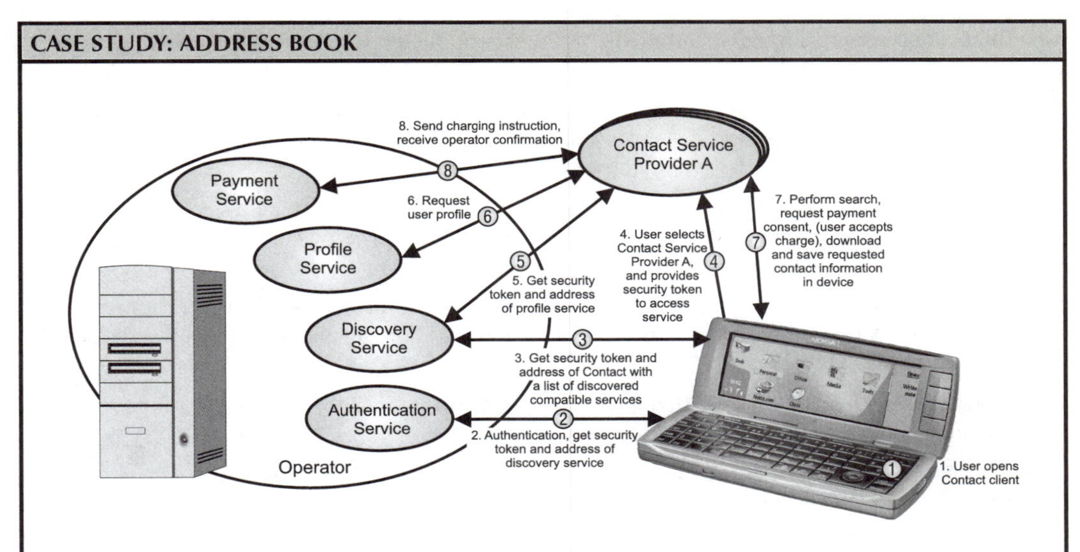

Figure 6-2. User-friendly application for address book search.

The above scenario illustrates how a mobile operator can set up a federated ecosystem with various service providers and enterprises to offer a multitude of services. One key component here is a Web service client that can interact with various back-ends. We demonstrate how a simple address book Web service application provides a user-friendly experience while utilizing complex services provided by different back-end servers. The advantages of creating such a system using the Liberty ID-WSF specifications are obvious: easy authentication, discovery, and service access.

For the enterprise use scenario at hand here, the operator provisions the clients on behalf of the enterprise and the different service providers. Standard protocols and field definitions are applied to enable interoperability with multiple back-ends. User participation is minimized – a key requirement for the frequent use of any client.

The use scenario phases are illustrated in Figure 6-2. The scenario covers the following phases:

1 **The user activates the client.**

2 **Authenticating client with an Authentication Service.** First, the client needs to be authenticated. This phase takes care of two tasks simultaneously: the client is provided with the address of the Discovery Service housing service descriptions, and it also receives a security token needed to access the Discovery Service. The level of authentication can be can be adjusted as required, ranging from completely automatic (our example) to the use of one-time passwords (not illustrated here). The response contains a security token and the URL of the Discovery Service.

3 **Getting a list of prospective service providers.** The message exchange between the client and the Discovery Service is also performed automatically. If the operator has deployed an Identity Federation Framework (using, for example, ID-FF 1.2 specifications for Single Sign-On), the Authentication Service can utilize that implementation and interact with the Identity Provider (in this example the mobile operator).

4 **Choosing a service provider (optional).** The user is presented with the available Contact Service Providers.

5 **Locating user's profile.** The selected Service Provider A contacts the Discovery Service to find out where the user's profile is stored. It is possible to add a request for user confirmation on sharing the required information, but this is not included in our example.

6 **Obtaining profile.** Service Provider A contacts the Profile Service to obtain user information – in this specific use case contact information. Nonetheless, it can be payment preferences, address information, phone number, and so on.

7 **Executing transaction.** The user performs the search, Service Provider A requests approval to charge the user's phone bill, the user accepts the charge, and the information is downloaded and stored in the device by the user.

8 **Charging the client.** Finally, Service Provider A instructs the operator to charge for the service by sending a request to the Payment Service.

It is possible to further extend the functionality in this use case by, for example, engaging a location service or resolving the user's location by other means. If the user is abroad, it may also be possible to offer local contact services for that country, in addition to service providers and the enterprise contact database located in the user's home country.

After the session, the user can access the operator portal (alternatively using the browser) to verify the amount charged by Service Provider A.

6.1 Liberty Identity Federation Framework (ID-FF)

The Liberty Identity Federation Framework (ID-FF) specification is a layered architecture with the most central features divided onto three layers. Starting from the top and proceeding down to lower layers, the central features are as follows:

- **Protocols** define the XML-based messages used to realize the features of the federated identity system. In practice, the messages define the data structures used to store and convey service data.
- **Profiles** define service messaging structures by taking the messages defined by protocols, and creating logical message exchange patterns by defining which messages are sent to whom and in what order.
- **Bindings** define how the profiles use lower-layer messaging and transport protocols.

6.1.1 Protocols

Protocols define the XML-based messages used to realize the Liberty architecture features, such as identity federation, SSO, and single logout. Each logical feature, such as SSO, has a protocol specification of its own. As the general description of the Liberty architecture above suggests, in the architectural stack, each protocol has a profile counterpart, which in a way provides a practical model for the usage of protocol messages.

The following sections introduce the key protocols and briefly describe the main elements of the protocol messages.

6.1.1.1 Single Sign-On and Federation

The Single Sign-On and Federation protocol realizes the SSO feature and identity federation.[2]

In the simplest SSO case, an SP requests the user's authentication assertion from an IDP by sending it an `AuthnRequest` message.[3] The message may contain parameters used to control the way the IDP handles the request. For example, the SP may require the IDP to confirm the operation from the user, or require the response to use a certain protocol profile (i.e., messaging/transport protocols).[4] In addition, it may be stated that a certain authentication context is necessary, requiring, for example, the use of an RSA SecurID passcode.

The IDP's response is sent in an `AuthnResponse` message, which contains the requested authentication assertions.[5] Alternatively, instead of providing the assertions (possibly over a non-secure connection), the IDP could send the SP an artifact referencing the actual assertion at the IDP. The SP could then use the artifact to fetch the assertion from the IDP (preferably over a secure connection).

The identities stored at an IDP can be federated in two ways by using the messages described above. If the SP initiates the process, the message exchange is almost identical to the assertion request described above. The only difference is that instead of asking for an authentication assertion, the SP requests the IDP to federate an identity stored at the IDP with the corresponding identity stored at the SP. The federation process itself does not require any initiative from the SP – IDPs can federate identities to SPs whenever necessary.

The single sign-on (SSO) protocol supports the presence of active intermediaries or so-called enhanced clients. These clients are able to receive an authentication request from a service provider and forward it to an identity provider, and then route the authentication response from an identity provider to the service provider. Enhanced clients can offer more efficient protocol bindings that do not rely on several HTTP redirect operations, thus producing improved performance in bandwidth-restricted scenarios. Enhanced clients also provide a higher degree of user control over the dissemination of the user's identity information in SSO, as SSO-aware software is incorporated into the user agent.

6.1.1.2 Single Logout

The Single Logout Protocol propagates information about a user's logout to all parties that have active connection sessions with the user.[6] Upon activation, the logout procedure is completely automatic, and it can be initiated at the IDP or the SP.

There are two possible scenarios that differ only slightly depending on whether the procedure was initiated at the SP or at the IDP. If the user activates the single logout feature at the IDP, the latter sends `<LogoutRequest>` messages to all SPs for which it has provided security assertions during the active session.[7] If the user activates the feature at the SP, the process is identical, except that the SP informing the IDP about the logout procedure does not receive a notification (it is already aware of the task).[8]

Due to the simplicity of this procedure, it either succeeds or fails. The response to the logout request is a `<LogoutResponse>` message, which provides status information on the task.[9]

6.1.1.3 Federation Termination Notification

The Federation Termination Notification Protocol ends identity federation between an IDP and an SP.[10] The protocol provides two use scenarios: one for the user and another for the SP. With the protocol, users can request an IDP to cease federating their identity with a certain SP. On the other hand, SPs can use the protocol to terminate federation by informing an IDP that the SP no longer accepts authentication assertions concerning a certain user.

The Federation Termination Notification protocol is based on an asynchronous message called `<FederationTerminationNotification>`.[11] The party activating the procedure sends this message to the other party. The message identifies both the user whose federation is to be terminated, and the sender of the message.

6.1.1.4 Name Registration Protocol

The Name Registration Protocol is the first of the two protocols not covered in the example scenario above.[12] Therefore, we begin with some background information, and then present the actual protocol messages.

When the initial account federation takes place between an IDP and an SP, the IDP creates an identifier called `IDPProvidedNameIdentifier` for the account owner. The IDP and the SP use this identifier to refer to the user. Alternatively, the SP can register an identifier of its own, called `SPProvidedNameIdentifier`, at the IDP. If both identifiers have been registered, then the IDP uses the `SPProvidedNameIdentifier` and the SP uses the `IDPProvidedNameIdentifier` to refer to the user in their mutual communication.

Thus, the Name Registration Protocol is used to manage name identifiers used by the IDP and the SP. After successful federation both parties refer to the user with the same identifier provided by the IDP (called `IDPProvidedNameIdentifier`). However, both parties can register new identifiers if necessary. With the Name Registration Protocol, an IDP can inform SPs about a new `IDPProvidedNameIdentifier`, and SPs can inform IDPs about a new `SPProvidedNameIdentifier`.

The Name Registration protocol is implemented using a message called `<RegisterNameIdentifierRequest>`.[13] The message provides the elements needed to register new identifiers and update old identifiers. When an IDP notifies an SP about a new identifier, it places the new identifier in an `IDPProvidedNameIdentifier` element and the old identifier in an `OldProvidedNameIdentifier` element.[14] When working in the opposite direction, an SP places the new identifier in an `SPProvidedNameIdentifier` element and the old identifier in an `OldProvidedNameIdentifier` element.[15]

Apart from status information, the procedure does not trigger any response messages to the initiator. The status response message, called `<RegisterNameIdentifierResponse>`, only contains a status code indicating the success or failure of the operation.[16]

6.1.1.5 Name Identifier Mapping

Like the previously described protocol, the last protocol covered in this section is not covered in our example scenario. The Name Identifier Mapping protocol is used in more complex communication scenarios taking place between an IDP and two SPs.[17] For example, let us consider a scenario where two SPs have separate federation links with an IDP. One of the SPs would like to contact the other concerning a user. This user has individual federations with both SPs.

The Name Identifier Mapping protocol allows one SP to communicate with another about a certain user for whom they do not share a federation. In other words, the first SP can contact an IDP to obtain a name identifier attached to the user, and then use this identifier to refer to the user in a message sent to the other SP. Thus, the two SPs do not need to share federation for the user.

The SP requesting the name identifier sends the IDP a `<NameIdentifierMappingRequest>` message.[18] The IDP replies with a `<NameIdentifierMappingResponse>` message, which contains the requested identifier, preferably in an encrypted form.[19] In this case, only the final recipient (i.e., the other SP) is able to decrypt the encrypted name identifier.

6.1.2 Profiles

A profile is a logical combination of ID-FF protocol messages and interactions taking place between different parties via transport protocols. In practice, profiles define the used protocol messages, messaging/transport protocols, and the exact message flow between different parties. Although it may appear complex, the actual realization of a profile is simple, as can be seen in the following protocol-specific sections.

It was noted in the beginning of the protocol section that each protocol has a profile counterpart with a similar name. In addition to the five protocol-counterpart profiles, this section describes two additional profiles used in the background, for example, to encrypt a `NameIdentifier`. The profiles are presented in the same order as the protocols, followed by the two extra profiles.

6.1.2.1 Single Sign-On and Federation

The Single Sign-On and Federation profiles are used to deliver a user's security assertions from an IDP to one or more SPs.[20] In addition, to mirror the features offered by the respective protocol, the profiles contain features used to federate identities.

The specification contains three profiles, each with distinct characteristics. Some of the concepts presented in the specifications, such as redirecting an HTTP form, can be realized with different technologies. Therefore, we do not discuss the details, and only present a high-level conceptual description.

The Liberty Artifact Profile is used to provide an artifact to the SP requesting an assertion. The SP uses this artifact to request the user's full assertion from the IDP.[21] The use of artifacts is explained in the Single Sign-On and Federation protocol.

With the Liberty POST Profile, the IDP can place an assertion in, for example, an HTML form sent to the client browser.[22] This form, and the embedded assertion, are then redirected to the SP either automatically or upon action initiated by the user.

The Liberty-Enabled Client and Proxy Profile allows active intermediaries to perform a role in SSO, forwarding requests to identity providers and responses to service providers. This profile is particularly useful in constrained bandwidth networks, and in environments where a high degree of user influence over the dissemination of user identity information is desired.[23]

6.1.2.2 Single Logout

Single Logout profiles are used to terminate all active user sessions created via a certain IDP.[24] The profiles transport the corresponding protocol messages between the user, the IDP, and the SP.

There are two types of Single Logout profiles: one based on HTTP redirect[25] and one based on SOAP/HTTP.[26] These translate into four practical scenarios, where the direction of the message flow differs depending on whether the procedure is initiated by the IDP or by the SP. (The IDP-initiated HTTP version also includes a profile that uses HTTP GET.[27])

6.1.2.3 Identity Federation Termination Notification

The Identity Federation Termination Notification profiles are used to provide transport-specific message-exchange patterns for federation termination protocol messages.[28] The profiles are used with the respective protocol messages to notify IDPs and SPs that the user wants to end identity federation between an IDP and an SP.

As in the case of Single Logout profiles, there are two types of Identity Federation Termination Notification profiles that are actually realized as four communication scenarios: one based on HTTP redirect and one based on SOAP/HTTP.[29] Again, the direction of the messages flow depends on the party initiating the procedure.

6.1.2.4 Register Name Identifier Profiles

The Register Name Identifier profiles enable the IDP and the SP to create name identifiers for clients with federated identities.[30] The identifiers can later also be modified using these profiles. The Register Name Identifier profiles are used to define practical message exchange patterns for the Name Registration Protocol.

As in the previously described profiles, in principle there are two types of name registration profiles, one based on HTTP redirect and on based on SOAP/HTTP.[31] In practice, however, there are four profiles due to the minor differences in the message exchange procedure depending on party initiating the procedure (the IDP or the SP). The difference is the direction of message flow.

6.1.2.5 NameIdentifier Mapping

The `NameIdentifier` Mapping Profile is related to the Name Identifier Mapping Protocol and the particular scenario in which an SP acquires a NameIdentifier that is used in communication with another SP regarding a user (see section 6.1.1.5 on the corresponding protocol).[32]

In contrast to the profiles described above, this specification contains only a simple request-response profile: the SOAP-based NameIdentifier Mapping profile.[33]

6.1.2.6 NameIdentifier Encryption

The `NameIdentifier` Encryption Profile is a single-purpose profile used to encrypt the NameIdentifiers during transport.[34] This makes it possible to shield the true identity of the owner of the identifier. The profile defines an XML Encryption-based method for encrypting the identifiers.[35]

This profile serves as an excellent example of the comprehensiveness of the security approach: the protected encryption placed in an `EncryptedNameIdentifer` element receives a different value for each request.[36] This makes it extremely difficult to link requests originating from a certain user with the actual user.

6.1.2.7 Identity Provider Introduction

Even though in this section we have only discussed scenarios with a single IDP, there is no limit to the number of IDPs in a federation system. Furthermore, as SPs may only want to cooperate with specific IDPs (e.g., due to business agreements), the SPs need a method that enables them to use the correct IDP with a certain client. To achieve this, the SPs can use Identity Provider Introduction profiles to locate the IDPs used by a client.[37]

The practical implementations of these optional profiles are solution-specific. All IDPs and SPs share a common domain, and a list of IDPs is placed in a Common Domain Cookie called _liberty_idp.[38]

6.1.3 Bindings

Bindings specify how the ID-FF protocol messages are conveyed via messaging and transport protocols. In practice, they define what information is placed in the different parts of the messaging protocol envelope, and how the envelope is transported (by using the features of a transport protocol).

The Liberty ID-FF Bindings and Profiles Specification describes the transport of data at a higher layer using SOAP (over HTTP).[39] In addition to addressing a set of general issues such as the use of SOAP header and body elements, the document provides header- and security solution-specific guidelines and restrictions on the transport of Liberty protocol messages via HTTP.

Authentication of communicating parties, message integrity, and confidentiality of communication are optional features, which are realized via secure low-level protocols. Available protocols include HTTP basic authentication, SSL3.0, and TLS 1.0.[40]

6.2 Liberty Identity Web Service Framework (ID-WSF)

The Liberty Identity Web Service Framework (ID-WSF) specifications enable the development of an identity service architecture based on Web services by using a layered SOAP architecture. When compared to the ID-FF specifications described above, the ID-WSF specifications do not adhere as strictly to uniform architectural style. Instead, the scope of each ID-WSF specification varies greatly: some affect only bindings while others define complete service sets by introducing new protocols and services.

ID-WSF is linked to the overall Liberty architecture in three ways:

- ID-WSF uses ID-FF identity federation services as a basis for its solutions that are based on Web services.
- ID-WSF provides new concepts and services for the architecture. For example, the Liberty Discovery Service is defined in ID-WSF.
- ID-WSF defines specifications that form a foundation for the ID-SIS specifications. One such specification is called Data Service Template, which is used by several ID-SIS specifications.

To illustrate the importance of the features offered by ID-WSF, we begin with an example. The scenario is somewhat more complex than the respective ID-FF case, although we are still dealing with a basic situation.

The following example illustrates the message exchange required to provide a user with a horoscope. The familiar players in the scenario are the user of a mobile device, the horoscope SP, and an IDP providing SSO features. The new ID-WSF actors are a Liberty Discovery Service that locates user identity information and an identity service (IS) that provides the actual identity information.

Needless to say, the first task is to realize the login procedure. In the example, we assume that the IDP service is provided by the cellular operator, and that user identification is handled by the UMTS Subscriber Identity Module (USIM) in the mobile device. Thus, once the user has entered her PIN to activate the USIM, and the mobile device has registered itself into the network, all services belonging to the operator's circle of trust are at the user's disposal.

The service session begins with an SSO procedure. First, the user contacts the horoscope SP. To complete its task, the SP needs the user's date of birth. Therefore, the SP wants to contact an IS hosting the user's personal profile (for a description of the Liberty ID-SIS Personal Profile, see section 6.3). To locate the identity service, and to retrieve the security assertions needed to obtain the information, the SP contacts a Liberty Discovery Service.

The Liberty Discovery Service (described in the ID-WSF Discovery Service Specification) provides the SP with a URL to the identity service, enabling the SP to request the user's date of birth. If needed, the Liberty Discovery Service can also provide the credentials required to access the identity service. Upon receiving the request, the identity service first ensures that the request is valid by prompting the user to approve the handing out of the information (this is realized using an ID-WSF Interaction Service). This ad-hoc questionnaire could also be used to update information or to request new information from the user. Once the request has been approved, the identity service delivers the user's date of birth to the horoscope service, which then provides the user with a personalized horoscope.

As noted above, the ID-WSF specifications do not have a uniform, logical structure. As a result, the order of the following document-specific sections might not appear as logical as in the ID-FF section. We begin with the specifications defining actual services, and continue with general documents affecting multiple areas (e.g., the security specifications). We conclude the section on ID-WSF by presenting the relevant bindings used to transport ID-WSF and ID-SIS data.

6.2.1 Authentication Service and Single Sign-On Service Specification

We begin the description of ID-WSF with the specification used to realize SSO and authentication features in the ID-WSF area. The Liberty ID-WSF Authentication Service and Single Sign-On Service Specification forms the core of authentication and SSO features based on Web services.[41] The specification affects a broad area, defining one high-level protocol and two services used to realize authentication and SSO features. Specified techniques include a general authentication protocol, an IDP-hosted authentication service, and a method to get authentication assertions needed to access ID-FF-based and other SPs.

6.2.1.1 Authentication Protocol

The ID-WSF Authentication Protocol defines a general method that enables authentication between any two Liberty ID nodes by using SOAP messaging.[42] This high-level protocol describes the exchange of authentication messages while hiding the actual security implementations and relying on SASL-defined methods to handle the details.[43] A successful exchange of authentication messages results in the authentication of the Web service consumer (the protocol also enables the authentication of the Web service provider).

6.2.1.2 Authentication Service

The ID-WSF Authentication Service serves two roles in a Web service-based authentication scenario: first, it enables Web service consumers (WSCs) to authenticate themselves to services that grant security assertions, and second, it enables WSCs to get security tokens needed to access other services.[44] The Authentication Service uses the ID-WSF Authentication Protocol.

6.2.1.3 Single Sign-On Service

The ID-WSF Single Sign-On Service provides WSCs with a method to acquire authentication assertions and security tokens needed to access services offered by SPs (including SPs not based on ID-WSF).[45]

6.2.2 Discovery Service

The Liberty ID-WSF Discovery Service Specification defines a directory service that is able to provide IS information and the security tokens needed to access the services.[46] The idea is to have a source for IS information able to provide the required security credentials on the go. These characteristics clearly differentiate the Discovery Service from UDDI, which focuses on providing service and SP information.

The Discovery Service and the information it handles are covered in more detail in section 5.1.10.

6.2.3 Interaction Service

The Liberty ID-WSF Interaction Service Specification defines methods for implementing an ad-hoc questionnaire which an IS can use to update the identity information it hosts or to get permissions to share the same information.[47] This query/update takes place informally in the middle of usual service functions.

In the example described in the beginning of section 6.2, this concept was presented at a general level. There, the IS wanted to verify that it should provide the requester with the user's date of birth. Next, we present a slightly more detailed example scenario.

In addition to the approval questionnaire described above, the IS can use a similar questionnaire to update the information it hosts. Let us consider a scenario with three participants: a user, an SP, and a Web service provider (WSP). The user is accessing a service provided by the

SP. The SP relies on the WSP to get the user's identity information. During the session, the WSP wants to update its database regarding the user's identity information. The update can be realized with the Interaction Service in the following way:

1. The WSP requests the SP to redirect the user session from the SP straight to the WSP.
2. The WSP updates the information by communicating directly with the user.
3. The WSP redirects the active session back to the SP (for more detailed examples, see the specification).[48]

In order to facilitate the above process, the specification defines a new element, a new protocol, and a new service.

6.2.3.1 UserInteraction Element

The `UserInteraction` element is a SOAP header used to advertise and control redirection and inquiry capabilities.[49] The sender of the element (in the example the SP functioning as a WSC) can request the WSP, for example, to request missing information from a user whenever necessary. It is also possible to prohibit all interaction with the user. Optional attributes can be used, for example, to indicate the languages that the user understands. This information is useful when constructing forms for questionnaires.

6.2.3.2 RedirectRequest Protocol

The `RedirectRequest` Protocol, and the profile defined in the protocol, define the processing rules and message exchange patterns for realizing interaction between a WSP and a user.[50] The protocol defines a new element called `RedirectRequest`, which is used to activate the actual redirect feature.[51] When the WSP wants to contact the user directly, it sends an SP a `RedirectRequest` element containing the URL to which it wants the user to connect. The accompanying profile defines the actual message exchange and the interaction between the WSC and the WSP by using, for example, the `UserInteraction` and `RedirectRequest` elements.

6.2.3.3 Interaction Service

The Interaction Service enables WSPs to request the interaction service to interact with the user in order to request information regarding a task.[52] In practice, the interaction service can dynamically create a form with the request, and the form is presented to the client to be completed by the user. When the requested information has been received, the interaction service delivers the information to the WSP. The specification defines an `InteractionRequest` element for requests and an `InteractionResponse` element for replies.[53]

6.2.4 Security Mechanisms

In contrast to ID-FF, which does not define any security specifications of its own, ID-WSF includes a security specification used to secure ID-WSF and ID-SIS services. The Liberty ID-WSF Security Mechanisms specification defines secure methods for the use of identity-based Web services.[54] Various other specifications in the architecture use it to realize their security requirements. In addition to the specification, the collection of documents contains several papers on security, covering issues such as trust models and privacy.[55]

The security mechanisms defined in the specification aim to cover a plethora of security requirements, such confidentiality and authentication. Chapter 5 provides an introduction to security issues related to Web services. This section briefly describes the features of the specification from a technical and feature-specific perspective. Security policies are not covered here, even though they do play a major role in defining the actual security solutions. Chapter 4 discusses the role of policies in organizations, and provides an example of a security policy.

We have divided our discussion of security mechanisms into the following two areas: confidentiality and privacy, and authentication. Both areas affect the communicating parties as well as the messages sent between them. Note that this approach touches only a small portion of the security concept. For a more comprehensive discussion, see the relevant specifications and summaries.[56]

6.2.4.1 Confidentiality and Privacy

Confidentiality of communication involves protecting not only the transport and the messages, but also the true identities of communicating parties. The Security Mechanisms specification defines a set of requirements for all three issues, and provides a set of possible solutions. The suggested protocols are not the only alternatives. Indeed, there are several ways to fulfill the requirements stated in the specification.

The first task is to secure the transport layer.[57] A single-hop link can be secured by using SSL or TLS, although from the complete end-to-end perspective, the use of these protocols precludes all intermediaries. Although the specification currently recommends using a set of TLS techniques, other security protocols may also be used, as long as the achieved security level equals the recommended solution.

Even with the transport layer secured (at least per-hop), the Web services architecture – which is filled with intermediaries – requires also the actual message to be secured.[58] In addition to the obvious need to encrypt the SOAP body, it is also necessary to protect the SOAP header information, because intermediaries might be able to manipulate the headers of the relayed messages.

Several ID-WSF Web service messages use identifiers that refer a client, a server, or a resource. The identifiers might disclose sensitive information simply by associating a client with a service or a resource. If a message exchange involves confidential information, which is almost certainly the case with all identity services, the privacy of the message exchange must be ensured by encrypting the identifiers.

6.2.4.2 Authentication

The Security Mechanisms specification considers the authentication of both the communicating entities and the messages exchanged between the entities. The used authentication mechanisms are defined with a set of URIs.[59] For example, a completely non-secure communication method is specified with the following URI: `urn:liberty:security:2003-08:null:null`.

The communicating entities can be authenticated in two ways: by authenticating only the message receiver or both the sender and the receiver.[60] In both cases, the authentication is realized using SSL/TLS. The following example presents a general case, where the message recipient and the message are authenticated using a combination of TLS and X.509: `urn:liberty:security:2003-08:TLS:X509`.

The Security Mechanisms specification defines message authentication by using X.590 v3 Certificate Message Authentication, SAML Assertion Message Authentication, or Bearer Token Authentication.[61]

6.2.5 Client Profiles

The Liberty ID-WSF Client Profiles Specification defines profiles for three sample categories of Liberty-Enabled User Agents and Devices (LUADs).[62] A LUAD is a user agent or device that participates in service message exchange within ID-WSF or ID-FF. The profiles address issues that LUADs typically encounter in the example scenario.

6.2.6 Reverse HTTP Binding for SOAP

The Liberty Reverse HTTP Binding for SOAP Specification (informally abbreviated as PAOS) defines a method by which an ordinary HTTP-based client (e.g., a Web browser) can receive SOAP requests via HTTP and is consequently able to provide SOAP-based services.[63] The scenario is simple: the client sends an HTTP request to a service, and receives a SOAP request within the HTTP response. In practice, this means that the client can function as an SP, even though it is not addressable and does not have server software as such.

The client advertises its PAOS support by adding a PAOS HTTP header to the HTTP requests it sends.[64] The interaction with the entity is described in message exchange patterns (for a practical example, see section 5.1.12).[65]

6.2.7 SOAP Binding

The Liberty ID-WSF SOAP Binding Specification defines the methods used to convey ID-WSF and ID-SIS service messages using SOAP (v1.1).[66] Like the respective ID-FF binding specification, SOAP binding defines how service messages are placed in SOAP message bodies and headers.

Additionally, the specification defines features missing from SOAP, and the headers used to realize these features. For example, one feature enables the association of messaging-specific requests and responses – a feature which is vital for client-server/request-response environments. The header element enabling this correlation of messages is called `Correlation`, and it contains, for example, a unique message-specific ID and a timestamp.[67]

6.2.8 Data Services Template

The Liberty ID-WSF Data Services Template Specification defines general protocols for querying and modifying data, and guidelines for data models.[68] It provides an abstract basis to be used by concrete data service instances. For example, the services described in the ID-SIS specification are based on the Data Services Template.

In addition to basic data query and modification features, the specification (v2.0) outlines subscription procedures that can be used to notify clients about changes in data, and to update the client's data set accordingly.[69] The update can be activated each time the data is changed or according to predefined triggers.

6.3 Liberty Identity Service Interface Specifications (ID-SIS)

The Liberty Identity Service Interface Specification set (ID-SIS) defines a group of identity Web services, each of which pertains to a category of information related to the user. At the time of writing, the available specifications can be used to handle personal information covering both business and private life. The ID-SIS Personal Profile Service can be used to store, for example, a person's name, address, messaging contact information, and a public encryption key.[70] The ID-SIS Employee Profile Service stores work-related information, such as employee ID, hiring date, and job title.[71] In addition to these two, the Liberty Alliance is working on other specifications, such as the ID-SIS Contact Book Service.

6.4 Summary

With this chapter focusing on the Liberty Alliance, we conclude the general section of our book. The preceding material has presented several basic and advanced Web service technologies.

Starting from chapter 7, the rest of this book describes the SOA for S60 platform and the Nokia Web Services Framework (NWSF). These platforms allow application developers to easily create identity-based or non-identity-based Web service clients for Nokia mobile devices. Both platforms provide security features used to access secure services that are protected, for example, with HTTP Basic Authentication. The introductory part concludes with a summary and a set of next steps for interested readers (in chapter 8). Chapters 9 and 10 are dedicated to Java and C++ application development examples that demonstrate in practice the concepts and platform features discussed in the first eight chapters.

Chapter 7: Enabling Mobile Web Services

This chapter presents two solutions that enable the development of mobile Web service clients for Nokia mobile devices. On the S60, the mobile Web services software component is called the SOA for S60 platform, and the Series 80 counterpart is called the Nokia Web Services Framework for Series 80 (NWSF).

The SOA for S60 platform will be available as a native platform component starting with S60 3rd Edition. A downloadable version will be available for devices based on S60 2nd Edition. Initially, the implementation supports C++ applications, with tools made available through Forum Nokia (www.forum.nokia.com). Java support for S60 will be implemented in future releases.

NWSF, with the corresponding functionality, supports C++ applications in the Nokia 9300 Smartphone and the Nokia 9500 Communicator. A downloadable version with support for Java applications is also available. Tools for both C++ and Java developers are available through Forum Nokia.

FORUM NOKIA
Forum Nokia membership consists of more than 2 million developers, and membership is free of charge. Forum Nokia Pro is a premium membership category with several additional benefits, such as business development support (events, co-marketing and PR support) and technical support (access to developer platform roadmaps, early access to pre-release tools, training, and prototype devices). Forum Nokia Pro also includes focused Zones, such as the Forum Nokia Pro Enterprise Zone and the Forum Nokia Pro Symbian Zone.

The SOA for S60 platform and the NWSF SDKs are two sets of libraries that allow mobile application developers to write service-oriented applications for Nokia mobile devices. These applications include those based on SOAP, which are commonly referred to as Web services. The extensible architectures of the SOA for S60 platform and NWSF allow for the addition of any future service-oriented technology portfolio.

The SOA for S60 platform and NWSF libraries contain application programming interfaces (APIs) in Symbian C++ for S60 and Series 80. In addition, Series 80 supports Java. The APIs simplify the development of service-consuming applications and, in the future, service-providing applications.

Although there are obvious differences in the underlying APIs, from a feature-specific perspective, the SOA for S60 platform and NWSF are identical. In the following sections introducing the two mobile Web service software components, we only use the term *SOA for S60 platform*, even though the discussion also applies to NWSF. However, the code examples presented in this chapter only apply to the SOA for S60 platform. The corresponding Series 80 C++ examples can be found in chapter 10, while the Java examples are presented in chapter 9.

Note that the original internal code name of the SOA for S60 platform and NWSF was Serene. The legacy of the name is that it, or the shorter version Sen, is visible in numerous library class names, and some SDK documents refer to the package with these unofficial names.

Figure 7-1. Nokia E61 is one of the devices equipped with S60 3rd edition.

7.1 Mobile Web Services Software Development Environment

The mobile domain is becoming open to the increasing availability of information and services on the Web. Mobile devices can browse the Web and are often connected to Personal Information Management (PIM) services (such as e-mail, scheduling, and contacts) along with other enterprise systems. There is demand to incorporate new applications and functions that make particular sense to mobile users, such as providing traffic and airport data, or other localized information.

Given the staggering number of mobile devices in use, and their upgrade rate, the opportunities are great for all parties involved (service consumers, enterprises, network operators, and service providers). Given also the dynamic nature of the mobile industry, the modular, loosely coupled nature of Web services technology promotes new business partnerships and collaboration opportunities for those managing mobile user accounts (such as mobile network operators) and different service providers.

As Web services are built on the foundation of the Web, offering mobile Web services is dependent on the physical capabilities of the mobile Web. The mobile Web differs from the fixed Web in significant ways. Mobile devices are subject to well-documented limitations. The processing power of mobile handsets may be restricted in terms of CPU capabilities, addressable memory, and persistent memory.

The user interface is limited in terms of both screen size and input facilities. The data network is constrained in both data throughput and latency. However, in all of these areas, mobile devices are improving dramatically.

Mobile operating systems and user interfaces are becoming increasingly powerful and sophisticated, as proven by the Symbian operating system and the Nokia S60 platform. These advanced platforms have opened up their local resources to host third-party applications. Current wireless packet switching protocols (such as GPRS) offer greater capacity, bit rates, and always connected functionality. These emerging mobile enhancements are creating positive conditions for mobile device–hosted Web service applications.

The constraints described above could limit the performance of Web services on the mobile device. Web service messages are sent as XML documents, so bandwidth, processing, and memory limitations may affect perceived application performance.

There are differences between developing software for a mobile device and developing software for a personal computer or a server. Because of the hardware and network limitations described above, it must be ensured that users are affected as little as possible by these limitations. This often means a great deal of work for the application developer.

The SOA for S60 platform has been designed to consider these differences and to ease the life of a Web services software developer in a number of ways. It does this by taking advantage of a number of features present in the Symbian operating system that are custom-built for developing software for the mobile device. These features are presented in the following sections.

7.1.1 Client-Server Architecture

The S60 and Series 80 platforms, which are both based on Symbian, provide a client-server architecture, which is used extensively in the SOA for S60 platform. A client process (the application) exchanges messages with Symbian server processes, managed by the Symbian kernel process. The benefit of this architecture is that if a server process crashes, Symbian takes care of the cleanup to prevent the whole platform from crashing.

In an environment where an application depends heavily on external network connections, separating the client application from the application server making the network connection allows the client to be somewhat insulated from network errors and latency. In the SOA for S60 platform, Web service consumer applications are also Symbian client applications. The SOA for S60 platform *Service Manager* is both a

Symbian server process and a Symbian client process of other Symbian server processes. This architecture ensures a level of robustness in the unstable networking and battery-powered environment of a mobile device.

7.1.2 Active Objects

The Symbian environment uses a feature called *Active Objects* to allow a software application to register a callback method. This means that the application can send a message to another process asynchronously, without having to wait for a response. This enables the client application to send a message and then perform other functions (such as updating a screen) rather than being forced to wait for a response.

The SOA for S60 platform makes use of active objects in many places, the most notable being the basic interface (`CSenServiceConsumer`) implemented as service consuming software applications. Active objects allow the application developer to write applications that can send Web service messages asynchronously. The SOA for S60 platform ensures that the consuming application is notified when a response message arrives, rather than having the application wait for a response from the Web service.

7.1.3 Symbian HTTP Stack

The Symbian platform offers access to network transport via an HTTP client stack. This feature allows, for example, HTTP Basic Authentication, and is the basis for the network connections established by the SOA for S60 platform.

7.1.4 XML Processing

The Symbian Series 80 and S60 platforms provide basic XML processing capabilities. The SOA for S60 platform supplements these by offering a simplified method of XML parsing to be used in Web services. These features enable efficient implementation of XML parsing on a mobile device, specifically for the purpose of parsing (typically) small XML messages (see section 7.7 below).

As noted earlier, the SOA for S60 platform APIs will be available in both Symbian C++ and Java. The Java environment for mobile devices also has to account for the differences inherent in mobile software development. This leads to the Java SOA for S60 platform architecture being based on a *Java Native Interface (JNI)* wrapper for Symbian SOA for S60 platform libraries. This allows the implementation of a suitable Java interface, while enjoying the excellent architecture for mobile software development provided by the Symbian environment.

The appropriate wrapper for the set of platform libraries allows a similar programming model to be provided across all languages that Nokia supports on its mobile devices.

As noted earlier in the book, Web service transactions involve a software application taking the role of either a service consumer or a service provider. Applications using SOA for S60 platform APIs may (in the future) assume either role, exposing mobile device-based functionality (such as your personal calendar or contacts) via a Web service interface. By using Web services, rather than Web pages (via a Web browser), software developers can create a better mobile-specific user interface because they are not tied to HTML. Web service messages tend to be small compared to HTML pages, as they do not contain presentation layer information.

7.2 SOA for S60 Platform Architecture

The architecture implemented in the C++ SDK is illustrated in Figure 7-2 below. Note that the figure only illustrates the C++ perspective. As the Java support of the SOA for S60 platform is based on the Service Development API, the corresponding Java illustration would be quite similar. The NWSF Java architecture is covered in more detail in chapter 9.

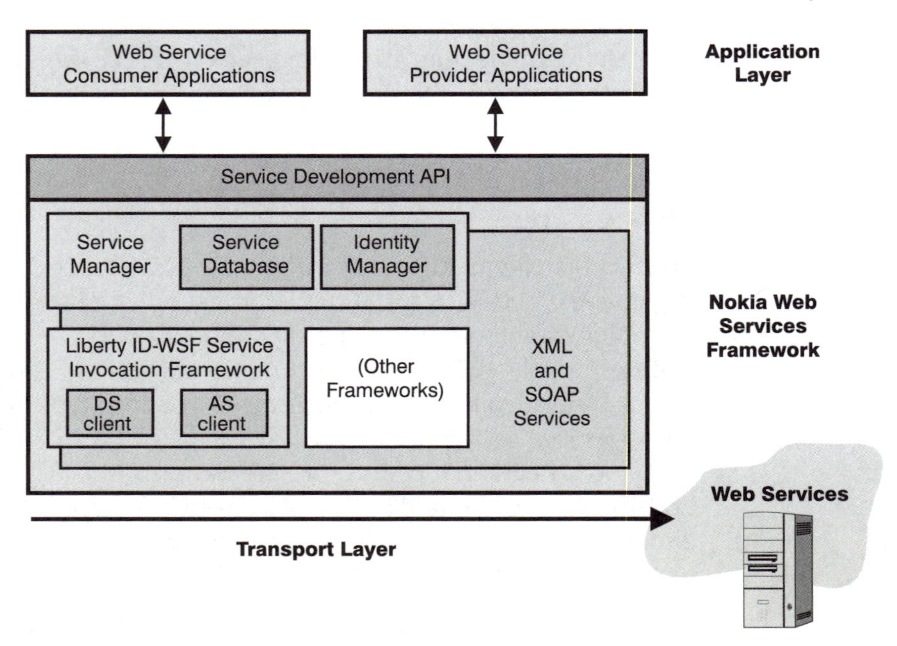

Figure 7-2. The SOA for S60 platform architecture.

Figure 7-2 presents an overview of the SOA for S60 platform architecture. Starting at the top, the *Web Service Consumer (WSC)* and *Web Service Provider (WSP)* applications are considered to be the clients in the framework. They use the Service Development API to access the functionalities provided by the SOA for S60 platform.

The SOA for S60 platform Service Manager is responsible for coordinating connections to services, and for storing identity and service information. The Liberty ID-WSF framework plug-in provides internal client (Web service consumer) applications that are able to connect to a Liberty Discovery Service and a Liberty Authentication Service, in addition to adding the framework-specific SOAP headers to submitted messages. Additionally, the SOA for S60 platform provides APIs for basic XML parsing, the creation of SOAP messages, and WS-Security headers.

The SOA for S60 platform Service Manager runs as a Symbian server application in its own Symbian process space. This provides a robust overall system, where the failure (crashing) of a client application process should not affect other running client applications, or the Service Manager process itself.

Currently, the SOA for S60 platform supports the development of service consumer applications that access Web services. The applications can be used to access many of the popular Web services available today, such as those offered by Google, eBay, and Amazon. Future versions will support the development and running of Web service provider applications by using this same architecture. This is covered in more detail below in section 7.5.

7.2.1 Frameworks

The SOA for S60 platform acts as *middleware*, located between the application developer and the networking and XML processing features available in the native mobile device platform.

It is often useful to write modular code, allowing each component of a system to be responsible for a specific part of overall processing while functioning relatively independently of other components. SOAs are based on this concept: systems are loosely coupled to each other, and well-defined interfaces specify the interaction between systems and their components.

There are potentially several SOAs, which could take advantage of these features, and could act in similar ways, requiring access to the same features.

For example, many services require authentication. A common framework for authentication could thus be used across several services. Also other features of systems (such as the way they handle identity) could be delegated to independent components, leaving a software application to handle only application-specific processing.

To cater to other potential service-oriented frameworks, the SOA for S60 platform offers a pluggable interface that allows such frameworks to be added to the SOA for S60 platform.

A framework plug-in implements all functionality needed to establish a connection to the type of service supported by the framework. The SOA for S60 platform Service Manager asks installed frameworks to provide it with a connection to the service requested by the client application, and the framework is then expected to execute all necessary tasks in establishing that connection.

Frameworks are also responsible for the wrapping and unwrapping of messages into framework-specific envelopes. For example, the ID-WSF framework plug-in wraps an XML application message into a SOAP envelope, and adds the required ID-WSF-specified SOAP headers to the message before dispatching it over a transport protocol to the requested service.

The first framework plug-in available for the SOA for S60 platform implements the Liberty ID-WSF framework for identity Web services. The SOA for S60 platform ID-WSF framework plug-in utilizes the Liberty ID-WSF SOAP Binding Specification and the Liberty ID-WSF Security Mechanisms Specification, which allow it to provide client applications with access to Liberty-compliant services. As part of this framework, the SOA for S60 platform provides two service consumer applications of its own: the *Discovery Service (DS)* client and the *Authentication Service (AS)* client. The SOA for S60 platform itself uses these clients to access the Liberty ID-WSF Discovery Service and the Liberty ID-WSF Authentication Service. This means that service consumers are both part of the framework, as well as client applications that use the framework.

The SOA for S60 platform supports the Liberty ID-WSF framework because it enables the support of Web services that exist in dynamic, complex business environments where services are modular, and authentication and other facilities are factored out of application services. In many cases, the interaction between clients and services demands more than just the basic Web service operations, as they concern issues such as identity and the configuration of authentication and security, which have been addressed in the Liberty ID-WSF specifications.

Although the basic APIs are quite simple, the SOA for S60 platform handles a number of underlying complexities to achieve the desired results, particularly in the process of establishing a connection to a service. This particular process is neatly handled by the service frameworks, which perform the often-complicated steps required in establishing the service connection. This functionality is described in more detail below.

All service consumer and provider applications act as SOA for S60 platform client applications. They request service access from the SOA for S60 platform Service Manager, which is then responsible for turning those requests into service sessions by using the available frameworks. The SOA for S60 platform maintains the service session to represent a session between one or more client applications and a particular instance of a Web service.

7.3 A Simple Example

The SOA for S60 platform SDK allows software developers to write Web service consumer applications by simply connecting to the Web service, sending XML application-specific messages to it, and processing XML application messages sent by the service. A simple example (a Web service called HelloWS) is presented below (Example 7-1). In the example, the Web service consumer makes a conversation request to a Web service provider, which replies with a "Hello Web Service Consumer!" response. The Symbian C++ code establishing the service connection is illustrated below:

```
void CConversationWS::ConstructL()
    {
    _LIT8(KServiceEndpoint, "http://example.nokia.com:8080/soap/
ConversationWS");
    CSenXmlServiceDescription* serviceDesc =
        CSenXmlServiceDescription::NewLC( KServiceEndpoint, KNullDesC8 );

    // KDefaultBasicWebServicesFrameworkID == "WS-I" as defined in
SenServiceConnection.h
    // This framework ID is used to invoke some service conforming with the
WS-I profile
    serviceDesc->SetFrameworkIdL(KDefaultBasicWebServicesFrameworkID);

    iConnection = CSenServiceConnection::NewL(*this, *serviceDesc);

    CleanupStack::PopAndDestroy(); // serviceDesc
    }
```

Example 7-1. Establishing a service connection.

The above code creates a simple description of a service. Typically, a service description identifies certain characteristics of the service, such as the following:

- **network endpoint**
- **schema contract** (the XML schema against which a message XML document can be validated, and possibly also a specification that governs the processing of conforming messages)
- **policy statements** that the service makes (e.g., regarding the use of a particular security token for message-level authentication, or the use of TLS for a secure connection)

In the simple example presented above, the service description is nothing more than a network endpoint. A service description can also be associated with a particular framework. Associating the service with a framework allows the SOA for S60 platform to decide which of its installed framework plug-ins can best provide the service connection. In the above example, the code explicitly states to the SOA for S60 platform which service framework should be used to handle messages sent to and received from this service. The SOA for S60 platform can then quickly return a connection handle to the calling application.

Once a connection has been established, the calling application can submit messages to the service. Example 7-2 below shows the application sending a message via the service connection obtained earlier:

```
TInt CAddressBookEngine::SendHelloL( const TDesC8& aRequestString )
    {
    // Create a new (empty) SOAP message to hold the application message
    CSenSoapMessage* soapRequest = CSenSoapMessage::NewL();
    CleanupStack::PushL(soapRequest);

    // Get handle to SOAP Body element
    CSenElement& body = soapRequest->BodyL();

    // Add new ConversationRequest element into SOAP Body
    CSenElement& helloRequest =  body.AddElementL(_L8("urn:nokia:ws:
samples"),

            _L8("ConversationRequest"));

    // Add child element which will hold actual request content
    CSenElement& helloString = helloRequest.AddElementL(_L8("HelloString"));
    helloString.SetContentL( aRequestString );

    // The iConverstationResponse holds the last response from the service
    delete iConversationResponse;
    iConversationResponse = NULL;

    // Submit the message to the service
    TInt retVal = iConnection->SubmitL(*soapRequest, iConversationResponse);
```

```
    LOG(*iConversationResponse);
    CleanupStack::PopAndDestroy(); // soapRequest

    // An intelligent application might be interested in checking
    // if retVal ==KErrSenSoapFault and parsing such response with
    // appropriate CSenSoapFault class to get details of what went wrong.
    return retVal;
    }
```

Example 7.2. Submitting a SOAP message to basic Web service.

The call to `iConnection->SubmitL(*soapRequest, iConversationResponse)` causes the SOA for S60 platform to send the message synchronously to the service and to wait for a response (this is known as a *blocking call*). The SOA for S60 platform provides APIs that allow asynchronous access to a Web service. The asynchronous mode allows the application to continue with other tasks after sending a message (for example, to display a progress bar), while the SOA for S60 platform waits for the response. Once the response has been received, the SOA for S60 platform notifies the consuming application about it. This callback is achieved through the Active Objects functionality provided by the Symbian platform, which requires an application to implement the methods of the `MSenServiceConsumer` interface (as this is descended from `CActive`, indicating that sub-classes are active objects) offered in the SOA for S60 platform SDK (`HandleMessage`, `HandleError`, and `SetStatus`).

An example application, such as the HelloWS application presented above, could produce the SOAP/HTTP message flow illustrated in Example 7-3 below:

1. The SOA for S60 platform WSC sends a request for a conversation with a WSP:

```
POST /soap/HelloWS HTTP/1.1
Host:
Accept: text/xml
Expect: 100-continue
User-Agent: Sen
Content-Length: 203
Content-Type: text/xml
SOAPAction: ""
<S:Envelope xmlns:S="http://schemas.xmlsoap.org/soap/envelope/">
<S:Body>
<ConversationRequest xmlns="urn:nokia:ws:samples">
<HelloString>Hello Web Service Provider!</HelloString>
</ConversationRequest>
</S:Body>
</S:Envelope>
```

2. HelloWS Web service responds:

```
HTTP/1.1 100 Continue
HTTP/1.1 200 OK
Date: Thu, 23 Sep 2004 19:57:39 GMT
Server: Jetty/4.2.21 (Windows 2000/5.0 x86 java/1.4.2_04)
Content-Type: text/xml
Content-Length: 270
<S:Envelope xmlns:S="http://schemas.xmlsoap.org/soap/envelope/">
<S:Body>
<ex:ConversationResponse xmlns:ex="urn:nokia:ws:samples">
<ex:ResponseString>
Hello Web Service Consumer!
</ex:ResponseString>
</ex:ConversationResponse>
</S:Body>
</S:Envelope>
```

Example 7-3. SOAP/HTTP message flow between the SOA for S60 platform and a Web service.

The above example was shown using Symbian C++. However, the NWSF Web service APIs are not restricted to C++. The NWSF APIs are already available to Java programmers, and access to Web services in the NWSF for Java API looks quite similar to that of the Symbian API. The Java code to access the HelloWS Web service is illustrated in Example 7-4 below. The APIs and the architecture of the Java solution are covered in more detail in chapter 9.

```
public void basicSample()
{
  try
  {
    WSFObjectFactory objectFactory = WSFObjectFactory.getInstance();
    ServiceManager serviceManager = objectFactory.getServiceManager();

    ServiceDescription serviceDesc =
      objectFactory.getServiceDescription(
        "http://example.nokia.com:8080/soap/ConversationWS",
        "urn:nokia:ws:samples",
        ServiceManager.BASIC_WS_FRAMEWORK_ID);

    ServiceConnection connection =
      serviceManager.getServiceConnection(serviceDesc, null);

    Element e = ElementFactory.getInstance().createElement(
                  null,
                  "ConversationRequest",
                  "urn:nokia:ws:samples");

    e.addElement("HelloString").setContent(
                                   "Hello Web Service Provider!");
    String helloRequest = e.toXml();

    // Alternatively request can be constructed directly as a string:
    //String helloRequest =
    // "<ConversationRequest xmlns=\"urn:nokia:ws:samples\">" +
    // "<HelloString>Hello Web Service Provider!</HelloString>" +
    // "</ConversationRequest>";
```

```
        System.out.println("Sending request " + helloRequest);
        String helloResponse = connection.submit(helloRequest);
        System.out.println("Response is " + helloResponse);
        connection.transactionCompleted();
    }
    catch (ElementException ee)
    {
        System.out.println("Exception: " + ee.toString());
    }
    catch (WSFException wsfe)
    {
        System.out.println("Exception: " + wsfe.toString());
    }
}
```

Example 7-4. NWSF for Java example.

7.3.1 Authentication for Basic Web Services

Even the most basic Web services may require client and/or server authentication, which can be implemented, for example, on the message level or the HTTP level (via HTTP Basic Authentication). The SOA for S60 platform provides features for creating Web service clients that can be authenticated directly by a Web service provider. In this process, the Web service authenticates a requester before allowing access to itself (this technique is known as self-authentication). This is different from Web services where authentication for the Web service can be accomplished by relying on an assertion from a third-party (the assertion stating that the requester was authenticated at a certain time, via a certain authentication method).

The SOA for S60 platform supports the addition of an OASIS Web Services Username Token to Web service messages, in addition to the support for authentication at the HTTP transport layer. The following code example shows how to provide HTTP Basic Authentication where the service itself authenticates the user at the HTTP layer:

```
void InitializeBasicConnection()
    {
    _LIT8(KServiceEndpoint,
        "http://example.nokia.com:8080/protected/ConversationWS");

    // Create a new identity provider, giving it identity details
    CSenIdentityProvider* idp =
        CSenIdentityProvider::NewLC( KServiceEndpoint);

    // This is a provider for some WS-I profile service
    idp->SetFrameworkIdL( KDefaultBasicWebServicesFrameworkID );

    // Set the credentials required for authentication
    _LIT8(KUser, "username");
    _LIT8(KPass, "password");
    idp->SetUserInfoL(KUser, KUser, KPass);

    // Register this identity provider to the SOA for S60 platform
    CSenServiceManager* serviceManager = CSenServiceManager::NewLC();
```

```
    serviceManager->RegisterIdentityProviderL(*idp);
    CleanupStack::PopAndDestroy(); // serviceManager

    // Note that this identity provider is identified by
    // the same network endpoint as that of the service,
    // which is protected with this authentication.
    // The identity provider is thus also the service provider.
    // Hence, connection may be initiated using idp as service
    // description.
    CSenXmlServiceDescription* serviceDesc = idp;
    iConnection = CSenServiceConnection::NewL(*this, *serviceDesc);
    CleanupStack::PopAndDestroy(); // idp

    // Now submit a message with above credentials
    // (username and password)
    TInt retVal = SendHelloL(_L8("Hello Web Service Provider!"));
    }
```

Example 7-5. Authentication using HTTP Basic Auth.

The above example involves the concept of an identity provider (IDP). In this case, the IDP is the entity providing the service. However, this may not always be the case. The IDP could be a party providing an authentication service on behalf of several different service providers. In this case, the service providers become relying parties for authentication assertions sent by the IDP, rather than directly performing authentication themselves. The code of the HTTP Basic Auth example results in a message flow similar to that of the following example:

1. *Initial request, posting the SOAP message.*

```
POST /protected/HelloWS HTTP/1.1
Host: example.nokia.com:8080
Accept: text/xml
Expect: 100-continue
User-Agent: Sen
Content-Length: 623
Content-Type: text/xml
SOAPAction: ""
<S:Envelope xmlns:S="http://schemas.xmlsoap.org/soap/envelope/">
<S:Header>
...
</S:Header>
<S:Body>
<ConversationRequest xmlns="urn:nokia:ws:samples">
<HelloString>Hello Web Service Provider!</HelloString>
</ConversationRequest>
</S:Body>
</S:Envelope>
```

2. The server responds, saying that the request was unauthorized and prompting for credentials.

```
HTTP/1.1 401 Unauthorized
WWW-Authenticate: Basic realm="example.nokia.com:8080"
Content-Type: text/xml
Date: Thu, 30 Sep 2004 16:59:15 GMT
Server: Jetty/4.2.21 (Windows 2000/5.0 x86 java/1.4.2_04)
<html>
<body>Unauthorized!</body>
</html>
```

3. The SOA for S60 platform adds the HTTP basic credentials to the request and re-sends.

```
POST /protected/HelloWS HTTP/1.1
Host: example.nokia.com:8080
Accept: text/xml
Authorization: Basic cm9sZTE6dG9tY2F0
Expect: 100-continue
User-Agent: Sen
Content-Length: 623
Content-Type: text/xml
SOAPAction: ""
<S:Envelope xmlns:S="http://schemas.xmlsoap.org/soap/envelope/">
<S:Header>
...
</S:Header>
<S:Body>
<ConversationRequest xmlns="urn:nokia:ws:samples">
<HelloString>Hello Web Service Provider!</HelloString>
</ConversationRequest>
</S:Body>
</S:Envelope>
```

4. The server authenticates the request and responds positively.

```
HTTP/1.1 100 Continue
HTTP/1.1 200 OK
Date: Thu, 30 Sep 2004 16:59:17 GMT
Server: Jetty/4.2.21 (Windows 2000/5.0 x86 java/1.4.2_04)
Content-Type: text/xml
Content-Length: 280
<S:Envelope xmlns:S="http://schemas.xmlsoap.org/soap/envelope/">
<S:Body>
<ex:ConversationResponse xmlns:ex="urn:nokia:ws:samples">
<ex:ResponseString>
Hello Web Service Consumer!
</ex:ResponseString>
</ex:ConversationResponse>
</S:Body>
</S:Envelope>
```

Example 7-6. HTTP Basic Auth message flow.

7.4 Identity Web Services Support

In addition to supporting basic Web service connections, the SOA for S60 platform offers simple APIs that provide service connections to Web services using the Liberty ID-WSF framework for identity Web services. When connecting to such Web services, more work is required from the SOA for S60 platform.

7.4.1 Connecting to an Identity Web Service

An identity Web service is one that provides a service on behalf of a certain identity, exposing attributes that are linked to that identity. Examples of such services include a geolocation service, which exposes the geographical location of the user on whose behalf the service is running, or a personal address card service (Liberty ID-SIS Personal Profile), which offers the user's address information via a Web service.

Identity Web services often involve major privacy issues. Because sensitive information is available via such services, authentication and confidentiality become very important. The service should establish a secure link between the authenticated requester and the identity on whose behalf the service is offered. It should be ensured that, for example, location or address information is not made available to unauthorized parties.

As noted previously, the Liberty ID-WSF provides specifications that allow the development of application services that do not need to worry explicitly about authentication. These specifications offer privacy-preserving features that services can use without having to build such features themselves.

Because of the privacy considerations discussed here and elsewhere in this book, accessing identity Web services can be complex. For example, it is probably not the case that a requester would discover an identity Web service merely by browsing a general Web services directory on the Internet. Even the discovery of identity Web services might be restricted to a very small group of authenticated requesters. In addition, it is quite likely that several identity Web services may be offered on the user's behalf, and therefore it makes sense to allow only certain requesters to access the full list of identity Web services. Thus, identity Web service discovery can offer a point of control both for the identity owner (the user) whose services are offered and for the service provider(s).

Because of these issues, the following steps are taken when connecting to an identity Web service:

1. Authenticate the service requester to an IDP, using an instance of a Liberty ID-WSF Authentication Service interface.
2. Once authenticated, discover the available services, and acquire security tokens that provide access to the requested service(s).
3. Connect to the service(s) previously discovered, supplying the acquired security tokens. (It is also possible to discover a service (e.g., via a UDDI service registry or by pre-provisioning the service consumer application) before being authenticated for it.)

The above steps are illustrated in the following figure:

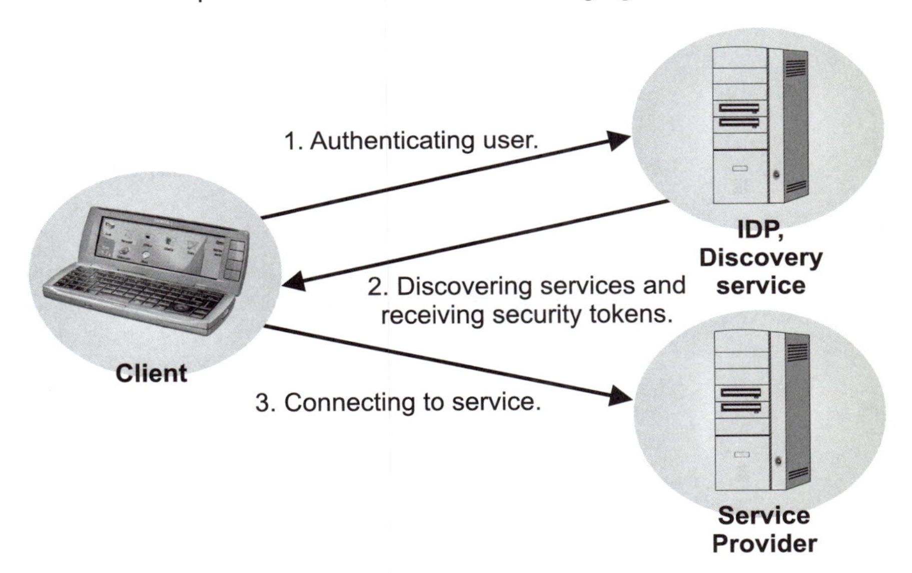

Figure 7-3. Interactions between the IDP, discovery service, and SP. In this particular example, the discovery service is provided by the IDP, although it may as well be a separate entity.

The SOA for S60 platform ID-WSF framework plug-in provides the functionality needed to create such connections via simple APIs. The above steps occur transparently to the application developer, who is responsible only for application-specific message formation and processing.

Now, let us see what the above steps look like in an XML message flow:

1. *Authenticate yourself in order to locate a discovery service.*

```
POST /soap/IDPAS HTTP/1.1
Host:
Accept: text/xml
Expect: 100-continue
User-Agent: Sen
Content-Length: 370
Content-Type: text/xml
SOAPAction: ""
<S:Envelope xmlns:S="http://schemas.xmlsoap.org/soap/envelope/">
<S:Header>
<sb:Correlation xmlns:sb="urn:liberty:sb:2003-08"
messageID="C8797D0D-9020-07FC-AF0A-5622C01F4A61"
timestamp="2004-09-27T14:50:45Z"/>
</S:Header>
<S:Body>
<sa:SASLRequest xmlns:sa="urn:liberty:sa:2004-04"
mechanism="PLAIN ANONYMOUS CRAM-MD5"
advisoryAuthnID="012345678901234"/>
</S:Body>
</S:Envelope>
```

2. *Response from the Authentication Service (AS), asking the SOA for S60 platform to continue with SASL PLAIN authentication.*

```
HTTP/1.1 100 Continue
HTTP/1.1 200 OK
Date: Mon, 27 Sep 2004 18:50:59 GMT
Server: Jetty/4.2.21 (Windows 2000/5.0 x86 java/1.4.2_04)
Content-Type: text/xml
Content-Length: 466
<S:Envelope xmlns:S="http://schemas.xmlsoap.org/soap/envelope/">
<S:Header>
<sb:Correlation s:mustUnderstand="1"
xmlns:sb="urn:liberty:sb:2003-08"
id="thisCorrHdr.2345"
messageID="i48b4353f50aca14
94665d61b93498c885449c868"
refToMessageID="C8797D0D-9020-07FC-AF0A-5622C01F4A61"
timestamp="2004-02-03T22:12:27Z"/>
</S:Header>
<S:Body>
<SASLResponse serverMechanism="PLAIN"
xmlns="urn:liberty:sa:2004-04">
<Status code="continue"/>
</SASLResponse>
</S:Body>
</S:Envelope>
```

3. *Continuing interaction with the AS.*

```
POST /soap/IDPAS HTTP/1.1
Host:
Accept: text/xml
Expect: 100-continue
User-Agent: Sen
Content-Length: 455
Content-Type: text/xml
SOAPAction: ""
<S:Envelope xmlns:S="http://schemas.xmlsoap.org/soap/envelope/">
```

```
<S:Header>
<sb:Correlation xmlns:sb="urn:liberty:sb:2003-08"
messageID="D93D6E95-A1F0-4A50-7938-1D1FA9D77918"
refToMessageID="i48b4353f50aca1494665d61b93498c885449c868"
timestamp="2004-09-27T14:51:00Z"/>
</S:Header>
<S:Body>
<sa:SASLRequest xmlns:sa="urn:liberty:sa:2004-04"
mechanism="PLAIN">
<sa:Data>ADAxMjM0NTY3ODkwMTIzNAAwMTIzNDU2Nzg5MDEyMzQ=</sa:Data>
</sa:SASLRequest>
</S:Body>
</S:Envelope>
```

4. The AS responds, accepting the authentication, and providing a resource offering for the discovery service (DS).

```
HTTP/1.1 100 Continue
HTTP/1.1 200 OK
Date: Mon, 27 Sep 2004 19:01:51 GMT
Server: Jetty/4.2.21 (Windows 2000/5.0 x86 java/1.4.2_04)
Content-Type: text/xml
Content-Length: 1169
<S:Envelope xmlns:S="http://schemas.xmlsoap.org/soap/envelope/">
<S:Header>
<sb:Correlation s:mustUnderstand="1"
xmlns:sb="urn:liberty:sb:2003-08"
id="thisCorrHdr.3456"
messageID="e44b8753f05abb1499657e61b83378c775219a768"
refToMessageID="D93D6E95-A1F0-4A50-7938-1D1FA9D77918"
timestamp="2004-02-03T22:12:27Z" />
</S:Header>
<S:Body>
<SASLResponse serverMechanism="PLAIN"
xmlns="urn:liberty:sa:2004-04">
<Status code="sa:OK"/>
<ResourceOffering entryID="1">
<ResourceID>
http://example.nokia.com:8080/soap/012345678901234
</ResourceID>
<ServiceInstance>
<ServiceType>urn:liberty:disco:2003-08</ServiceType>
<ProviderID>http://example.nokia.com:8080/soap/</ProviderID>
<Description>
<CredentialRef>2sxJu9g/vvLG9sAN9bKp/8q0NKU=</CredentialRef>
<SecurityMechID>
urn:liberty:security:2004-04:null:Bearer
</SecurityMechID>
<Endpoint>http://example.nokia.com:8080/soap/IDPDS</Endpoint>
</Description></ServiceInstance>
</ResourceOffering>
<Credentials notOnOrAfter="2004-09-28T18:28:44Z">
<saml:Assertion xmlns:saml="urn:oasis:names:tc:SAML:1.0:assertion"
AssertionID="2sxJu9g/vvLG9sAN9bKp/8q0NKU=">
...
</saml:Assertion>
</Credentials>
</SASLResponse>
</S:Body>
</S:Envelope>
```

5. The SOA for S60 platform now asks the DS whether it knows the service type "urn: nokia:ws:samples".

```
POST /soap/IDPDS HTTP/1.1
Host:
Accept: text/xml
Expect: 100-continue
User-Agent: Sen
Content-Length: 679
Content-Type: text/xml
SOAPAction: ""
<S:Envelope xmlns:S="http://schemas.xmlsoap.org/soap/envelope/">
<S:Header>
<sb:Correlation xmlns:sb="urn:liberty:sb:2003-08"
messageID="F5683B9F-DF73-AA9E-D01A-82CFFE8F6341"
timestamp="2004-09-27T14:51:01Z"/>
<wsse:Security xmlns:wsse="http://schemas.xmlsoap.org/ws/2003/06/
secext">
<saml:Assertion xmlns:saml="urn:oasis:names:tc:SAML:1.0:assertion"
AssertionID="2sxJu9g/vvLG9sAN9bKp/8q0NKU=">
...
</saml:Assertion>
</wsse:Security>
</S:Header>
<S:Body>
<Query xmlns="urn:liberty:disco:2003-08">
<ResourceID>
http://example.nokia.com:8080/soap/012345678901234
</ResourceID>
<RequestedServiceType>
<ServiceType>urn:nokia:ws:samples</ServiceType>
</RequestedServiceType>
</Query>
</S:Body>
</S:Envelope>
```

6. The DS knows the service type, and provides the information and credentials to allow access to an instance of that service.

```
HTTP/1.1 100 Continue
HTTP/1.1 200 OK
Date: Mon, 27 Sep 2004 19:01:53 GMT
Server: Jetty/4.2.21 (Windows 2000/5.0 x86 java/1.4.2_04)
Content-Type: text/xml
Content-Length: 1066
<S:Envelope xmlns:S="http://schemas.xmlsoap.org/soap/envelope/">
<S:Body>
<QueryResponse xmlns="urn:liberty:disco:2003-08">
<Status code="ok"/>
<ResourceOffering entryID="1">
<ResourceID>
http://example.nokia.com:8080/soap/ConversationWS/5678901234
</ResourceID>
<ServiceInstance>
<ServiceType>urn:nokia:ws:samples</ServiceType>
<ProviderID>http://example.nokia.com:8080/soap/</ProviderID>
<Description>
<CredentialRef>2sxJu9g/vvLG9sAN9bKp/8q0NKU=</CredentialRef>
<SecurityMechID>urn:liberty:security:2004-04:null:Bearer
</SecurityMechID>
<Endpoint>
http://example.nokia.com:8080/soap/ConversationWS
</Endpoint>
</Description>
</ServiceInstance>
```

```
</ResourceOffering>
<Credentials notOnOrAfter="2004-09-28T18:28:44Z">
<wsse:BinarySecurityToken EncodingType="wsse:Base64Binary"
wsu:Id="ia1575535f5b0712dbff7033db0721e4f838390f3"
xmlns:wsse="http://schemas.xmlsoap.org/ws/2003/06/secext"
xmlns:wsu="http://schemas.xmlsoap.org/ws/2003/06/utility">
AZoOuAM4BdMxtKugmt1qiwZze11vQb/m5udOPOTa8Y5L
</wsse:BinarySecurityToken>
</Credentials>
</QueryResponse>
</S:Body>
</S:Envelope>
```

7. The SOA for S60 platform WSC sends a "Hello" request.

```
POST /soap/HelloWS HTTP/1.1
Host:
Accept: text/xml
Expect: 100-continue
User-Agent: Sen
Content-Length: 528
Content-Type: text/xml
SOAPAction: ""
<S:Envelope xmlns:S="http://schemas.xmlsoap.org/soap/envelope/">
<S:Header>
<wsse:Security xmlns:wsse="http://schemas.xmlsoap.org/ws/2003/06/
secext">
<wsse:BinarySecurityToken
EncodingType="wsse:Base64Binary"
wsu:Id="ia1575535f5b0712dbff7033db0721e4f838390f3"
xmlns:wsse="http://schemas.xmlsoap.org/ws/2003/06/secext"
xmlns:wsu="http://schemas.xmlsoap.org/ws/2003/06/utility">
AZoOuAM4BdMxtKugmt1qiwZze11vQb/m5udOPOTa8Y5L
</wsse:BinarySecurityToken>
</wsse:Security>
</S:Header>
<S:Body>
<ConversationRequest xmlns="urn:nokia:ws:samples">
<HelloString>Hello Web Service Provider!</HelloString>
</ConversationRequest>
</S:Body>
</S:Envelope>
```

8. HelloWS Web service responds.

```
HTTP/1.1 100 Continue
HTTP/1.1 200 OK
Date: Thu, 23 Sep 2004 19:02:39 GMT
Server: Jetty/4.2.21 (Windows 2000/5.0 x86 java/1.4.2_04)
Content-Type: text/xml
Content-Length: 270
<S:Envelope xmlns:S="http://schemas.xmlsoap.org/soap/envelope/">
<S:Body>
<ex:ConversationResponse xmlns:ex="urn:nokia:ws:samples">
<ex:ResponseString>
Hello Web Service Consumer!
</ex:ResponseString>
</ex:ConversationResponse>
</S:Body>
</S:Envelope>
```

Example 7-7. Example message flow using an identity service.

A message flow like this is generated by simply changing a few lines of the basic code presented in Example 7-1. The changes are shown below:

```
_LIT8(KServiceContract8, "urn:nokia:ws:samples");
// Liberty ID-WSF service connection may be established
// with contract only, assuming that required identity provider
// has been registered to the SOA for S60 platform
// (endpoint and credentials to some ID-WSF authentication service).
iConnection = CSenServiceConnection::NewL(*this, KServiceContract8);
```

Example 7-8. Authentication using an identity service.

The main change here is that instead of using a network endpoint to identify the service description to the SOA for S60 platform, a service contract is used. This contract identifies the service as one of type `urn:nokia:ws:samples`.

The HelloWS application is not the best example of an identity Web service, but it illustrates the differences between the basic service approach and the identity Web service approach. All differences occur at the SOA for S60 platform middleware level, and they are almost invisible to the application developer. The next section provides more information about how the middleware assists the application developer.

7.4.2 SOA for S60 Platform Service Database

In order to enable the access described above, the SOA for S60 platform maintains a database of service information, which it consults every time a connection is requested by a consuming application. The service information comprises service connection endpoint information and possible credentials used to access the service.

When the SOA for S60 platform is asked for a connection, the SOA for S60 platform Service Manager searches the service database for existing information on the service. If a service is description is located, and it either needs no credential or is described with a valid credential, then the SOA for S60 platform can immediately provide a connection to the requesting application. However, if the service is not described completely (perhaps missing the network endpoint, or a credential), then the SOA for S60 platform performs additional tasks to acquire and register the complete service description it needs in order to provide the requester with a valid connection.

In XML, the service information entry for the HelloWS service could look like this:

```
<ServiceDescription framework="ID-WSF">
<Contract>urn:nokia:ws:samples</Contract>
<Endpoint>http://example.nokia.com:8080/soap/ConversationWS</Endpoint>
</ServiceDescription>
```

Example 7-9. HelloWS service information.

The third-party authentication service used by the HelloWS service can also be described in the service database, along with other services.

By using the `RegisterServiceDescription` API described in the following section, an application developer writing a client for a particular Web service can provision the service database of the SOA for S60 platform with some or all of the details required to access the Web service. The application developer can also de-register a service description.

7.4.3 Service Association

Now, let us imagine that the description of ConversationWS looks like this:

```
<ServiceDescription framework="ID-WSF">
<Contract>urn:nokia:ws:samples</Contract>
<TrustAnchor>http://example.nokia.com:8080/soap/</TrustAnchor>
</ServiceDescription>
```

Example 7-10. ConversationWS service information with a service association.

This would indicate that the HelloWS service trusts assertions issued by the provider identity described in the `TrustAnchor` element. If the SOA for S60 platform already had a description in the service database that listed a service with a provider ID corresponding to a provider ID denoted by HelloWS as trusted, then the SOA for S60 platform would be able to contact that IDP to obtain the authentication required to access the HelloWS service. Note also that in the above service description, the network endpoint for HelloWS is missing. However, since the SOA for S60 platform can contact the trusted authentication service for HelloWS, it is likely that this authentication service will notify the SOA for S60 platform about a discovery service from which to obtain the necessary information in order to access the HelloWS service. That sequence of events is described in the XML message flow illustrated in Example 7-7 above.

The SOA for S60 platform provides simple API facilities that allow the registration of a service into the SOA for S60 platform. In the case of a Liberty-compliant identity Web service, this registration can be a partial description. In case a Web service connection is created without utilizing a framework, the description should contain at least a network endpoint, because this scenario contains no framework-provided discovery mechanism that could obtain the remaining information required to contact the service. The example below illustrates the use of this API to register an identity-providing service and the service provider itself, and then to associate the two:

```
_LIT8(KServiceId8, "");
_LIT8(KProviderID8, "");
serviceMgr->RegisterIdentityProviderL( *idp );
serviceMgr->RegisterServiceDescriptionL( *serviceDesc );
// Associate service provider with identity-providing service.
serviceMgr->AssociateServiceL(KServiceId8, KProviderID8 );
```

Example 7-11. Associating a service provider with an identity-providing service.

7.4.4 Identity Providers

An IDP is an entity responsible for maintaining user accounts. IDPs can authenticate users for which they have accounts. They may also provide authentication assertions regarding these users to other system entities, such as other Web service providers. As noted above, IDPs can be registered into the SOA for S60 platform and associated with other services, creating a trust link between the service and the IDP. In addition to the service database, the SOA for S60 platform maintains a database of identity information. In the database, the SOA for S60 platform can store user account details for identity providers. In this case, when a user requests a service connection that uses a particular IDP, the SOA for S60 platform provides the user account information to the IDP during authentication. Sometimes this information is entered into the database at the time a consuming application is installed in the system for a particular service. Registering user account information into the SOA for S60 platform is simple, as demonstrated in the following example:

```
CSenIdentityProvider* idp =
    CSenIdentityProvider::NewLC(KASEndpoint8,  KASContract8);
idp->SetProviderID(KASProviderId8);
idp->SetUserInfoL(KNullDesC8, KTestAdvisoryUserId8, KTestPassword8);
```

Example 7-12. Registering user account information.

The IDP can then be registered normally into the SOA for S60 platform.

7.4.5 Service Matching

Several actual service instances of a particular service type may exist at the same time. For example, a user could operate a location Web service running on the mobile device, and another somewhere on the fixed Internet. As a requester, the user may wish to choose manually between service instances of the same type, or may wish to simply provide a pattern to the SOA for S60 platform, and have the SOA for S60 platform determine the best service from which to obtain the results. The SOA for S60 platform will always attempt to match the network endpoint and/or the service contract (if these details are specified). In addition, the SOA for S60 platform enables the specification of *pattern facets*, which are particular characteristics of a service that allow the best of a group of service instances to be selected. For example, let us imagine that one service instance charges a fee for access whereas another instance of the same service type can be accessed free of charge. Facets are specified as URNs, and can be supplied in a service pattern, prompting the SOA for S60 platform to attempt to locate the best service match based on the supplied facets. API functions are provided to specify the facets to the SOA for S60 platform.

A service may support particular options (or facets) that are of interest to a service consumer. For example, some Liberty-compliant services may offer the consumer the option of sending full XPath queries for data to the service. As with service policies, a consumer may wish to work with a particular service (within a class of services) because it supports a particular facet.

The SOA for S60 platform provides a service matching API allowing one to create a connection to a service based on the facets of that service. This is part of the ServiceDescription API. One example of ensuring that a particular service has a specific facet enabled is shown below. The example allows the requester to ask for a connection to a Liberty ID-SIS Geolocation Service, which returns "high accuracy" location results:

```
CSenXmlServiceDescription* pSD = CSenXmlServiceDescription::NewLC() ;

//note on the service pattern that the "accuracy:high" facet should be set

CSenFacet* pAccuracy = CSenFacet::NewL() ;
CleanupStack::PushL(pAccuracy) ;
pAccuracy->SetNameL(KAccuracyHighFacetName) ;
pSD->SetFacetL(*pAccuracy) ;

// create a service connection to this service in the normal way
```

Example 7-13. Using a facet as a search pattern for a service connection.

7.4.6 Service Policies

A service may require certain policies to be observed by a requester accessing the service. For example, a service may require that requesters always use TLS for connections to the service, and that they supply a WS-Security SAML token to provide message-level authentication.

A service can indicate that it has a particular policy in a number of ways. One way is to indicate the policy in a response message to the service consumer – a very dynamic way of indicating a policy.

Other places where policy might be indicated by a service include the WSDL description of the service, or its registration details in a discovery registry (such as a Liberty Discovery Service or a UDDI registry).

A service-consuming application may also wish to indicate to the SOA for S60 platform that it requires certain policies to be observed. For example, a particular service consumer application may indicate that it requires all outgoing network connections to be made over a particular secured transport (for example, requesting that a VPN-secured internet access point be used on the mobile device).

As both the service-providing application and the service-consuming application may have policies – even conflicting policies – it is sometimes necessary to perform policy matching. If the service consumer wants to use a particular type of service, and some instances of that service have policies that do not match the policies of the consuming application, then the consumer can choose to use a service whose policies do match its own, instead of overriding its own policies.

Such policies are handled by the SOA for S60 platform at the framework level. For example, in the Liberty ID-WSF framework plug-in, security policies are registered with a discovery service. When the SOA for S60 platform discovers a Web service, it also discovers and employs the policies registered at the discovery service to aid it in establishing connections to the service.

7.5 Web Service Providers

This chapter has concentrated on the development of Web service consumer applications. These applications use one or more Web service providers to offer their application features. However, it will also be possible to run Web service provider applications with the SOA for S60 platform. In this case, requesters are able to contact a Web service running on a mobile device. For example, PAOS (see section 5.1.12) allows a personal profile service to run on the mobile device, supplying profile information (such as the date of birth) to a Web site and allowing the Web site to customize a Web page based on the provided identity-related content.

Exposing mobile device–based functionality, such as one's personal mobile calendar, enables new service applications to make use of this data. For example, imagine that a GPS-capable mobile device exposes the coordinates of the device via a service interface. These coordinates could be delivered to a network-based service providing the location of the nearest coffee shop.

The SOA for S60 platform SDK will provide an API that allows the creation of Web services accessible from the mobile device. This so-called *hostlet* API allows an application developer to utilize existing Symbian client APIs (for example, to use calendar information from the mobile device calendar, or to access data from the GPS device) and expose this data via a Web service interface. A simple test service, using these APIs is shown below:

```
const TDesC8& CSenTestProvider::ServiceContractUri()
    {
    return (_L8("urn:com.nokia.provider:dummy")) ;
    }
TInt CSenTestProvider::ServiceL(const TDesC8& aContract,
                                           MSenHostletRequest& aRequest,
                                 MSenHostletResponse& aResponse)
    {
     TInt retVal(KErrNone);
    _LIT8(KBody,
        "<nn:TestResponse
xmlns:nn=\"urn:com.nokia.test\">Success!</nn:TestResponse>");
    aResponse.SetResponseBodyUtf8L(KBody);
    return retVal;
    }
```

Example 7-14. Simple implementation of a Web service provider.

7.6 SOA for S60 Platform as a Pluggable Architecture

Web Services is not the only service-oriented technology that exists today (for example, it is possible to send non-SOAP XML messages). To cater for any future SOA technology developments, the SOA for S60 platform offers a pluggable middleware architecture, which can be expanded with plug-ins to support other service-oriented frameworks. It is possible to install a new framework via the Symbian ECOM plug-in architecture, and we can anticipate the development of new framework plug-ins that will be added to the SOA for S60 platform, providing access to other types of services. One such framework plug-in, the Liberty ID-WSF framework plug-in, was introduced in the beginning of this chapter.

7.7 SOA for S60 Platform XML Processing

Web service messages are delivered as XML documents. These documents may have been produced according to an XML schema, and may be validated against that schema to ensure that the document meets at least basic syntax checking. With apologies to those readers who are familiar with the basics of XML processing, and with the intent of introducing the XML processing capabilities provided by the SOA for S60 platform, we begin with an introduction to parsing Web service messages.

7.7.1 Introduction to Processing XML Messages

Very often, Web service messages contain a SOAP `Envelope` XML element, which contains zero or more SOAP header blocks (inside a SOAP `Header` XML element), and a SOAP `Body` XML element. The header blocks and the body may be examined by several processing components in sequence.

Any part of a software application may have to a) check that input XML is valid for further processing and b) map the input XML document to its own internal representation (which may not resemble the XML input document at all).

In order to do any processing of a service message, one or more message components must be *parsed*. The parsing operation converts the input XML into a representation that a piece of software can utilize. For example, if the message contains a list of addresses, parsing that message could turn the XML shown below into a Java object representing the owner of the addresses:

```
<pp:PP xmlns:pp="urn:liberty:id-sis-pp:2003-08">

  <pp:AddressCard>
    <pp:AddressType>urn:...:home</pp:AddressType>
    <pp:Address>
      <pp:PostalAddress>99 Shady Street</pp:PostalAddress>
      <pp:PostalCode>TN320FB</pp:PostalCode>
    </pp:Address>
  </AddressCard>

  <pp:AddressCard>
    <pp:AddressType>urn:...:work</pp:AddressType>
    <pp:Address>
      <pp:PostalAddress>102 Shady Street</pp:PostalAddress>
      <pp:PostalCode>TN320FB</pp:PostalCode>
    </pp:Address>
  </AddressCard>

</pp:PP>
```

Example 7-15. Liberty ID-SIS Personal Profile–based address cards.

In the Java object, the addresses could be represented by the 'addresses' `ArrayList` in the following example:

```
Public class Person extends XMLObject
{

  Java.util.List addresses = new java.util.ArrayList() ;

  // constructor

  Public Person( )
  {…}

  Public void buildFrom( XMLDocument aPersonDocument )
  {
    // parse the input document into the properties of this object
  }

  Public void addAddressesFromXml( XMLDocument anAddressCollection )
  {
    // parse the input document--which looks like that shown above--
    // into the 'addresses' ArrayList() defined as a
    // property of this class
  }

  // other methods
}
```

Example 7-16. Parsing an XML document into a Java ArrayList.

The code shown above is an example of how one might translate an XML input representation into something used in software. There are quite a few others. There are also tools (such as JAXB by Sun Microsystems) for creating Java classes that are translated directly from an XML schema – the classes thus closely resemble the input XML schema structure. Not all software applications want to bind their XML input documents so tightly to their code structure.

In all cases, the incoming XML document is read by software and translated as something useful to the software application.

Parsing software turns the string representation delivered as an XML document into an internal data structure of some kind. In the case of XML, this internal representation is often a tree: each element in the document may contain child elements (in the example above, `Address` is a child element of `AddressCard`), and the child elements appear as branches. Each XML entity is called a *node* in XML parsing parlance, and the result of parsing an XML document in this manner is a node tree.

An XML parser that returns a node tree according to interfaces specified by the W3C DOM Working Group is known as a DOM parser (where DOM stands for Document Object Model).

As this method of parsing results in a tree of objects representing each XML entity in the input document, it can result in heavy memory usage when a large document is being processed.

An alternative parsing strategy is called SAX (Simple API for XML). This method of parsing does not depend on the building of an internal tree representation of an XML document. Instead, a SAX parser reports parsing events (such as *start element*, *end element* and so on) to the calling application. The calling application can then register a handler method for each parsing event that it is interested in. This allows a) the option of only writing code to process the parts of the XML that the user (i.e., the software application) is interested in, and b) to turn an XML document representation directly into the object model of the user's choice.

In Web service applications, different components of a software application are usually responsible for different parts of an input XML message. For example, one component may handle the WS-Security SOAP header block, while another extracts the content of the message body. Thus, each component examines a different part of the SOAP message. It is quite likely that application developers do not want an individual component to process the entire XML document in order to execute its tasks. Therefore, it looks like the SAX approach may be more efficient for parsing incoming messages. However, SAX can be complicated: registering a callback method for each XML event the user is interested in results in a complicated application execution that may be difficult to follow for many developers (and testers).

So far, we have only talked about the processing (parsing) of incoming XML messages. But are there ways to simplify the creation of XML messages, without forcing a developer to use plain strings of XML in their code?

In this case, one can either create methods of one's own that allow an object of some class to serialize itself as XML, or use the DOM API approach and create (subclasses of) DOM nodes and arrange them in a tree that knows how to serialize itself as XML.

7.7.2 SOA for S60 Platform XML API

As noted earlier, there are different ways to parse an XML document, and a software application has several different needs concerning an XML processor:

- Parsing into an appropriate structure from an XML document
- Ability to locate one or more elements from an XML document that the application is interested in
- Ability to provide classes that allow the hiding of the XML nature of an application object (such as the Person object shown above in Example 7-16)

The S60 platform provides general XML processing features of this nature. They support a full DOM and SAX parsing approach, and allow software developers to create their own XML node trees able to serialize themselves as XML. The features are based on a well-known libxml2 parser (for more information, visit <xmlsoft.org>).

In addition to the platform's parsing capabilities, the SOA for S60 platform API itself provides some further XML processing APIs. They are intended to simplify some of the aspects of developing XML-based software. The APIs are described in the following sections.

7.7.2.1 CSenBaseElement

The `CSenBaseElement` class (and descendents) can be used in a DOM-like fashion to deal with a node tree. This is particularly useful when creating a set (node tree) of objects to be serialized as an XML message.

The following example is taken from a service consumer application for the eBay auction Web services. (This example deliberately omits the typical Symbian memory management code (`CleanupStack::PushL() and Pop()`) for clarity.):

```
CSenElement* secHeader =
  CSenBaseElement::NewL(_L8("RequesterCredentials"));

        secHeader->SetNamespace(KNsUri);
        secHeader->AddAttrL(_L8("S:mustUnderstand"), _L8("0"));

        //eBayAuthToken
        CSenElement* element = CSenBaseElement::NewL(_L8("eBayAuthToken"));

        // set the content of the containing element
        // to be the auth token string

        element->SetContentL(KeBayAuthToken);

        // add the element to the parent element (ie.
        // as a child node of the tree)

        secHeader->AddElementL( *element );

        //Credentials
        element = CSenBaseElement::NewL(_L8("Credentials"));
        element->SetNamespace(KNsPrefix, KNsUri);

        //DevId
        CSenElement* childEl = CSenBaseElement::NewL(_L8("DevId"));
        childEl->SetContentL(KDevId);

        // add a child node to the Credentials element
        element->AddElementL( *childEl );

        //DevId 2
        childEl = CSenBaseElement::NewL(_L8("AppId"));
        childEl->SetContentL(KAppId);

        // add another child node to the Credentials element
```

```
        element->AddElementL( *childEl );

        //DevId 3
        childEl = CSenBaseElement::NewL(_L8("AuthCert"));
        childEl->SetContentL(KAuthCert);

        // add another child node to the Credentials element
        element->AddElementL( *childEl );

        //Add Credentials element as a child of RequesterCredentials
        secHeader->AddElementL( *element );
```

Example 7-17. Creating credentials for eBay.

The resulting object tree based on XML elements could then be serialized as an XML text fragment (for example, by calling `secHeader->AsXmlL()).`

7.7.2.2 CSenDomFragment

The `CSenDomFragment` class is useful for parsing a message where you know you are probably interested in the entire structure of the XML message being parsed, but do not necessarily know the name of the top-level element of the incoming message. Thus, you might not be able to register a SAX handler for the unknown element. This approach is most useful when you know that the XML representation of the data is practically the same as the object representation that you would like to use in your application, as in the following XML example:

```
<AddressCard>
  <FirstName>John</FirstName>
  <LastName>Doe</LastName>
  <Email>John.Doe@Nokia.Com</Email>
  <StreetAddress>
    <Address1>1666 Pudding Lane</Address1>
    <City>London</City>
    <Country>England</Country>
  </StreetAddress>
</AddressCard>
```

Example 7-18. An XML-based example address card.

Creating an object class such as `CAddressBookContact` equipped with better methods that return items corresponding to those in the XML example above is quite simple: sub-class the `CSenDomFragment` as shown in the following (incomplete) code fragment:

```
AdressbookContact.h

          // CONSTANTS
        _LIT8(KElementNameN,          "N");
        _LIT8(KElementNameGiven,      "GIVEN");
        _LIT8(KElementNameFamily,     "FAMILY");
        _LIT8(KElementNameEmail,      "EMAIL");
```

```
    // CLASS DECLARATION
       class CAdressbookContact: public CSenDomFragment
         {
         public: // Constructors
         static CAddressBookContact* NewL();

         // New functions
         // @return first (given) name of a person
         TPtrC8 FirstName();
         // @return last (family) name of a person
         TPtrC8 LastName();
         // @return the email address of a person
         TPtrC8 Email();
         /**
          *  Search method for content. Localname of
          *  parent and child element has to match.
          *
          *  @param aParent localname of parent which is supposed to
          *         own spesified child
          *  @param aChild localname of child which content is being
          *         requested
          *  @return the content of child element
          */
         TPtrC8 ContentoOfChild(const TDesC8& aParent,
                                const TDesC8& aChild)
         // more methods ...
              }

AdressbookContact.cpp

        TPtrC8 CAdressbookContact::FirstName()
          {
          return ContentoOfChild(KElementNameN, KElementNameGiven);
          }

      TPtrC8 CAdressbookContact::LastName()
          {
          return ChildValue(KElementNameN, KElementNameFamily);
          }

      TPtrC8 CAdressbookContact::Email()
          {
          return ContentOf(KElementNameEmail);
          }

     TPtrC8 CAdressbookContact::ContentoOfChild(const TDesC8&
aParent,
                                                      const TDesC8&
aChild)
          {
          CSenElement* pElement = AsElement().Element(aParent);
          if (pElement)
              {
              pElement = pElement->Element(aChild);
              if(pElement)
                  {
                  return pElement->Content();
                  }
              }
          return KNullDesC8();
          }
```

Example 7-19. Subclassing CsenDomFragment to create a new object class.

In order to further parse the `StreetAddress` element in the example, the `CAddressBookContact` class could override the default `StartElementL()` and `EndElementL()` methods of the `CSenDomFragment` in order to create an object of a `CStreetAddress` class responsible for further parsing that structure (if necessary).

The class might be instantiated by the following piece of code illustrating the use of the `CSenBaseFragment::StartElementL()` method:

```
void CAddressBookParser::StartElementL(const TDesC8& aNsUri,
                                       const TDesC8& aLocalName,
                                       const TDesC8& aQName,
                                       const RAttributeArray& aAttrs)
    {
    // If this is a contact card, parse it into CAddressBookContact
    // In most cases, namespace match is typically checked, too.
    if (aLocalName == KCard)
        {
        iErrorState = ENoError;
        CAddressBookContact* delegate = CAddressBookContact::NewL();
        CleanupStack::PushL(delegate);

        // Add a new contact to the list of contacts,
        // which will take the ownership.
        iContacts.Append(delegate);

        // Start parsing this fragment with newly
        // created CAddressBookContact instance.
        DelegateParsingL(*delegate);

        CleanupStack::Pop(delegate);
        return;
        }
    }
```

Example 7-20. Using delegation to pick fragments of XML stream.

IMPORTANT NOTE
Care should be taken with this approach, as a `CSenDomFragment` constructs an in-memory XML node tree, which may exhaust all available memory. For Web service messages, which are typically quite small with little internal structure, this is usually not a major concern.

7.7.2.3 CSenBaseFragment

The `CSenBaseFragment` class allows the developer to register an object that can be used to parse a particular XML entity present in an XML document. Using this class is like the traditional (and simpler) way to register a SAX event handler for the occurrence of the named XML entity. A developer may create sub-classes of this class to deal with the processing of specific XML entities (for example, the `CSenSoapMessage` utility class itself is descended ultimately from `CSenBaseFragment`). It is then possible to delegate parts of the XML parsing process to individual classes responsible for different parts of the structure of the incoming message. The particular advantage of this approach is memory saved in only parsing the structures needed by the application (and avoiding the wrapping structures required but not necessarily important to the application developer).

This approach is analogous to the traditional SAX parsing method. A sub-class of `CSenBaseFragment` may implement its own custom parsing routines (by overriding the `StartElementL()` and `EndElementL()` methods of the base class). The sub-class is then responsible for maintaining a state machine to track where the class is in the parsing process.

If an incoming message is known to contain several parts (or a particular class is only interested in a single part of the message) then the application developer can create a sub-class of `CSenBaseFragment` to parse the incoming message. Only the requested element (specified using local name, namespace, and `ns` prefix) is then saved by default (i.e., no child elements of the node). Implementing the `StartElementL()` and `EndElementL()` methods can then be used to create even more parsed nodes (possibly by creating other sub-classes of `CSenBaseFragment`).

As an example, let us consider the previous example where we used a `CSenDomFragment` to encapsulate a complete address card. In that case, the whole XML representation of an `AddressCard` was converted into an in-memory XML nodeset. If, however, the application were only interested in the `StreetAddress` portion of the XML, a `CSenBaseFragment` could be constructed passing the local name `StreetAddress`. This object could then be used to analyze the whole message, parsing only the `StreetAddress` portion once its start tag was found.

As a final note, it is possible to easily produce a `CSenBaseElement` representation of a fragment by using the `CSenBaseFragment->AsElement()` or `CSenBaseFragment->ExtractElement()` API methods to retrieve an element representation. The difference between these these two calls is that `ExtractElement()` actually transfers ownership of the memory resource associated with that element to the calling application – this can be important if the caller is, for example, expecting to control the garbage collection of that element.

7.7.3 Tips on Choosing a Parsing Strategy

Below are two key ideas related to choosing the appropriate parsing strategy for incoming XML messages:

1. Do not use a single `CSenDomFragment` to build a tree for a large XML document – this has a high impact on memory usage. It is better to think about the data that you actually need from a large document and extract it using a SAX-based approach (e.g., creating sub-classes of `CSenBaseFragments`, each with the name of the element you want to use from the input XML).
2. `CSenBaseFragment` (and `CSenDomFragment`) were intended to allow developers to create sub-classes that extend `CSenBaseFragment` and are responsible for parsing a named fragment of XML. These fragments are then managed by the software application in a way that corresponds to the software's internal object structure.

7.8 Writing Code for Web Service Consumer Applications

The SOA for S60 platform SDK provides the basic features needed to connect to Web services by using the simple strategy of first connecting to a service and then communicating with the service via XML messages. The process has been described in the examples above.

Many programmers will, however, want to use a higher-level interface to interact with Web services. They may want to take an object-oriented approach, and use object classes that hide the SOA for S60 platform connection. With a conversion tool available for Eclipse and Microsoft Visual Studio, it is possible to take a WSDL document and automatically generate client stub code and classes that represent the data types (defined in the WSDL document) to access the Web service. This may help further hide the complexity inherent in Web services, giving the appearance that a developer is interacting with objects locally, rather than having these objects generated from Web service descriptions and having their data filled by a connection to a remote Web service.

The advantage is that the developer would not have to write much of the actual code shown above, but would instead be able to take a Web service description document, run a translation tool, and have a set of object classes that represent the messages and types needed to access the Web service. The developer would then manipulate these object classes without having to see the underlying need to make Web service message requests, or the code needed to fulfill those requests.

One of these tools is the Nokia WSDL to C++ Wizard for the Series 80 Platform. It can be downloaded from Forum Nokia at http://www.forum.nokia.com/main/1,6566,034-961,00.html. A similar tool (as part of Nokia Enhancements for Java for Series 80 Platform) is available for generating Java code at the Forum Nokia Pro Web site: https://pro.forum.nokia.com/site/global/tech_resource/tools/tools_detail/l_java_enhancements.jsp.

The tools listed above were available at the time of writing; additional tools are expected to become available for the S60 platform.

7.9 Summary

This chapter has discussed the Nokia SOA for S60 platform and the NWSF Web services architecture and SDKs. These currently available technologies offer features that allow Java and Symbian C++ developers to access Web services of all kinds, ranging from the most basic services to complex environments employing third-party authentication or stringent privacy requirements—all from a Nokia mobile device. In addition, the SOA for S60 platform and NWSF offer a pluggable architecture that allows the development of frameworks to support other service-oriented applications than those already available today.

Chapter 8: Summary and Next Steps

This book has provided an introduction to mobile Web services, with a focus on the convergence of Web services developed for the fixed Internet with those developed for the mobile environment. This convergence means that the consumers and providers of services can interact with interoperable Web service messages from anywhere, by using mobile devices or fixed Internet clients and servers. From a business perspective, the advantage of a converged set of standards for interoperability is enormous. Today, these standards include SOAP messaging, description (WSDL), discovery (UDDI), and security (WS-Security), along with the WS-I Basic Profile. In this book, we have emphasized the importance of identity and identity-based Web services, and have discussed standards published by the Liberty Alliance (ID-FF and ID-WSF). Trust is a critical component of meaningful interaction. It is achieved by establishing identity through authentication, and it is used in authorizing access to personal information and identity-based services. The protection of privacy is a critical issue for consumer services and important in enterprise deployments as well. It is not enough to develop a variety of standards and then simply hope for the best. Instead, the protection of privacy must be ensured with an end-to-end system architecture, as was discussed with respect to the Liberty Alliance architecture.

We have also emphasized the importance of helping application developers to manage the complexities associated with service architectures, such as messaging, description, discovery, security, and the details related to privacy and identity-based Web services. We have described the SOA for S60 platform, which simplifies the use of service architectures by providing one or more frameworks. For example, the SOA for S60 Platform enables an application developer to create Web service consumers and providers for mobile devices. These consumers and providers can then use Web services and identity-based Web services with less effort.

We believe that the introduction of service-oriented technologies to mobile devices will create an entirely new range of mobile applications, including those discussed in this book. They will include consumer applications (arrange my travel itinerary with airlines and hotels), enterprise applications (connect me to the travel approval and expense systems), and government applications (register me to vote, process my tax payment, update my personal information). Mobile networks can provide mobile information, such as presence, location, and payment services to enhance the offerings of service providers (find me a restaurant near 33rd St and 5th Avenue). Similarly, a mobile device can offer services ranging from user interaction to sharing personal contact and calendar information.

Although there are many advantages to the convergence of the fixed and mobile Internet, it is important to be aware of the specific characteristics of the mobile environment. They include disconnected and intermittent connectivity, messaging costs, addressability issues, and user interface limitations. Understanding these issues makes it clear that identity-based services create added value (for example, no need to interact to repeatedly establish identity) and that a platform such as the SOA for S60 platform can enable application developers to avoid dealing with some of these complex issues, because they are handled by the platform.

This book is a starting point designed to introduce key concepts and raise awareness about opportunities. The next step is to visit the Forum Nokia Web site to learn more about how you can create new applications by using the SOA for S60 platform and the underlying frameworks.

Chapter 9: Java Client Development

Earlier in this book, we have explained the meaning of a SOA and the Web services technologies enabling the creation of one such service-oriented architecture. The last two chapters in this book concentrate on providing readers with instructions on how to develop mobile Web service clients for Nokia mobile devices. This chapter describes Java service development APIs and the underlying Service Development API, while the next chapter focuses on C++ development.

The Nokia Web Services Framework for Java (NWSF for Java) relies on the Service Development API, which is a native Symbian programming library distributed as NokiaWSF.sis. (This file accompanies the Nokia Enhancements for Java for Series 80 SDK.) The API of the Web Services Framework for Java is implemented by NWSF for Java. This chapter provides code examples that illustrate the discussed technical concepts.

NOTE
Please check the Forum Nokia Web site and the present book's Web page on the publisher's Web site (http://www.wiley.com) for source-code downloads.

9.1 Prerequisites

Since this chapter focuses only on Web service features implemented with Java on Nokia devices, reading it requires familiarity with Java programming, and especially with the J2ME environment. A working knowledge of Java Servlet technologies is also required for testing some of the examples. It is also assumed that you are familiar with the concepts related to Web services, and knowledge of the Liberty Alliance specifications (described in chapter 6) is certainly an asset. We recommend that you study the Developer's Guide to Nokia Web Services Framework for Java and Nokia Web Services Framework for Java API JavaDocs together with this material.[1]

TESTING THE EXAMPLES

At the time of writing, the first versions of the Nokia Web Services Framework (NWFS) have just been released, and our examples are based on the NWSF release for the Symbian Series 80 Platform 2.0. The NWSF APIs for the Series 60 Platform may vary in terms of minor nuances. However, even if you are working with a later release of NWSF while reading this book, the same interfaces and methods that are described here should apply, because Nokia assures binary compatibility across releases.

To test the sample applications discussed in this chapter, you need three sets of tools. First, in order to utilize the NWSF libraries, you need the base SDK for the Symbian (C++) environment, the *Series 80 Platform 2.0 SDK for Symbian OS*. Second, you need the *Nokia Prototype SDK 4.0 Beta for JME*. This SDK can be integrated with a variety of IDEs (such as Eclipse). Third, you need a toolset titled Nokia Enhancements for Java for Series 80 Platform. This is a software package containing two middleware components that enrich the Java APIs of Series 80 Platform devices. The components are Nokia Message Queue and Nokia Web Services Framework for Java.

9.2 Nokia Web Services Framework for Java Architecture

The NWSF for Java API is a set of Java interfaces and classes that allow application developers to write Web service applications. Currently support is provided for Web service consumer applications, but in the future, it will also be possible to write Web service provider applications that run on Nokia mobile devices.

DEVELOPMENT TOOLS

The examples in this book illustrate the capabilities and usage of the various interfaces and methods of the framework. To allow the reader to follow the examples presented in this chapter, the sample applications are kept relatively simple.

When creating Web service applications, developers expect to have access to tools that relieve them from tedious programming (i.e., typing) involved in serializing and deserializing data into SOAP-based XML streams and in handling connection management. Nokia provides tools that automatically generate the proxy code required for these tasks. These tools create stub files with methods that resemble the remote operations specified in the WSDL service definitions. The tools also enable the creation of auto-generate skeletons for the processing of received asynchronous messages. As a result, developers only need to enter the appropriate application logic in the skeletons.

Nokia's Web service tools are designed for the typical development process of a Web Service Consumer application (WSC). Such an application normally starts out with a definition and implementation of a Web service, which is then formalized as a Web service description (in WSDL). Various tools are available for creating these WSDL specifications, such as the Web Tools Platform for Eclipse and the Visual Studio tools for Microsoft Windows. Many service providers, such as eBay, Amazon.com, and Google also publish WSDL specifications for their services. Given these WSDL specifications and the associated XML schema definitions of data types, the Web service tools provided by Nokia can significantly reduce the amount of manual work carried out by programmers in converting XML-style definitions into programming code (both Java and Symbian C++).

The Web service tools are available as part of the SDK downloads, as described above (*Nokia Enhancements for Java for Series 80 Platform*). They can be either installed as plug-ins for IDEs (e.g., Eclipse) or executed on the command line. Similar tools are available for generating Symbian C++ proxy code.

The NWSF API provides a set of interfaces that allow developers to write service consumer applications based on SOAP, WS-Security, WS-I, and Liberty ID-WSF. In addition, some interfaces are provided to aid XML processing.

Figure 9-1 shows an overview of the NWSF architecture. Starting at the top, WSC and the future WSP applications are considered clients of the framework. They use the NWSF API to access the provided functionality. The functionality offered by the NWSF API is built on top of the underlying Service Development API. The Service Development API is described in more detail in chapter 7.

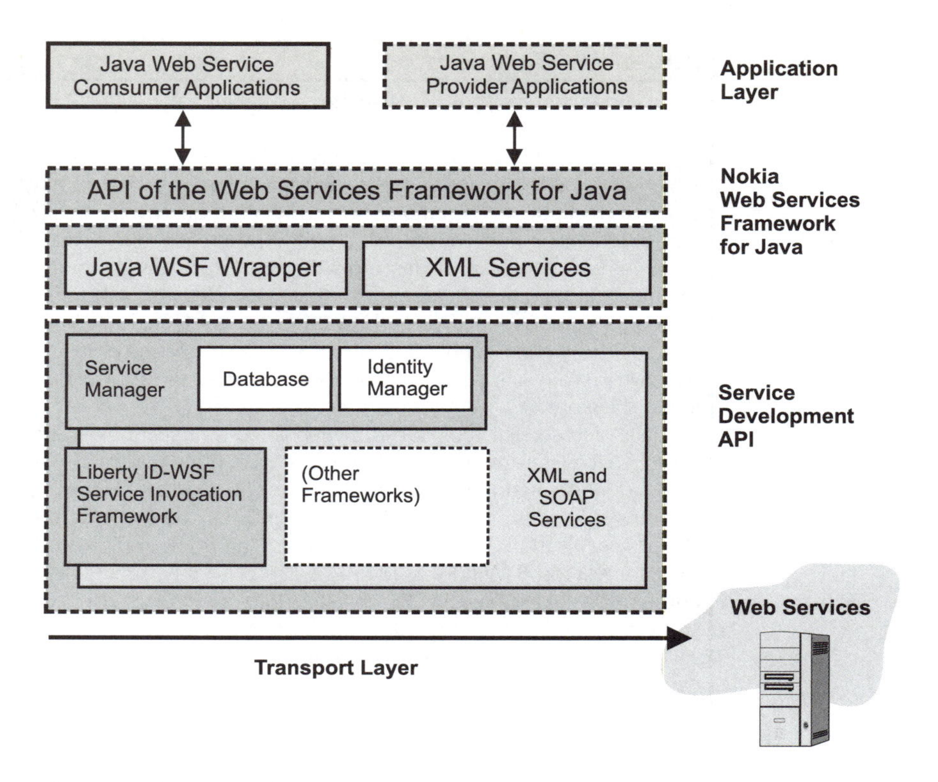

Figure 9-1. Overview of NWSF architecture.

9.2.1 General Structure of NWSF API

The NWSF API consists of a number of Java interfaces and classes. The main interfaces are structured as follows:

1. `com.nokia.mws.wsf.ServiceConnection` – This interface represents a connection with a Web service.
2. `com.nokia.mws.wsf.ServiceManager` – This is the main interface of NWSF for Java. It enables client applications to register services and IDP information, and to obtain connections to Web services.
3. `com.nokia.mws.wsf.ServiceDescription` – This interface is used to describe Web services. It contains information such as the contract URI of the service and the URL used to connect to the service.
4. `com.nokia.mws.wsf.IdentityProvider` – This interface represents IDPs that the framework uses to authenticate users. IDPs are services at which the user has an account. The IDPs know the

user and other services trust them to make assertions regarding the identity of the service-requesting user.

5. `com.nokia.mws.wsf.WSFObjectFactory` – This object contains factory methods for obtaining instances of `ServiceManager`, `ServiceDescription`, and `IdentityProvider` instances.
6. `com.nokia.mws.wsf.xml.Element` – This is a utility for basic XML processing. The interface can be used to ease the burden of XML-processing the protocol messages, because it provides a simple DOM-like interface to XML.

9.2.2 Configuration

The Service Development API is configured by XML files, which are located in the c:\system\data directory in the Symbian emulator environment. APIs are provided to allow the developer to interact with the configurations.

The first of these files (`SenSessions.xml`) contains the stored descriptions of services, and can end up looking, for example, as follows (initially it is empty):

```
<SenConfiguration xmlns="urn:com.nokia.Sen.config.1.0">

<Framework xmlns="urn:com.nokia.Sen.idwsf.config.1.0" class="com.nokia.Sen.
idwsf.IdentityBasedWebServicesFramework"/>

  <ServiceDescription framework="ID-WSF">
    <Contract>urn:liberty:as:2004-04</Contract>
    <Endpoint>http://192.168.11.2:8080/soap/IDPAS</Endpoint>
    <ProviderPolicy/>
    <ProviderID>http:// 192.168.11.2:8080/soap/</ProviderID>
  </ServiceDescription>

  <ServiceDescription framework="ID-WSF">
    <Contract>urn:liberty:disco:2003-08</Contract>
    <Endpoint>http:// 192.168.11.2:8080/soap/IDPDS</Endpoint>
    <Credentials> notOnOrAfter="2005-11-28T18:28:44Z" >
      <saml:Assertion xmlns:saml="urn:oasis:names:tc:SAML:1.0:assertion"
AssertionID="2sxJu9g/vvLG9sAN9bKp/8q0NKU=">
      ...
      </saml:Assertion>
    </Credentials>
    <ProviderPolicy/>
    <ProviderID>http:// 192.168.11.2:8080/soap/</ProviderID>
    <ResourceID>http:// 192.168.11.2:8080/soap/IDPDS/4b3d</ResourceID>
    <TrustAnchor>http:// 192.168.11.2:8080/soap/</TrustAnchor>
  </ServiceDescription>

  <ServiceDescription framework="ID-WSF">
    <Contract>urn:nokia:ws:samples</Contract>
    <Endpoint>http:// 192.168.11.2:8080/soap/HelloWS</Endpoint>
    <Credentials notOnOrAfter="2005-11-28T18:28:44Z">
      <saml:Assertion xmlns:saml="urn:oasis:names:tc:SAML:1.0:assertion"
AssertionID="2sxJu9g/vvLG9sAN9bKp/8q0NKU=">
      ...
      </saml:Assertion>
```

```
    </Credentials>
    <ProviderPolicy/>
    <ProviderID>http:// 192.168.11.2:8080/soap/</ProviderID>
    <ResourceID>http:// 192.168.11.2:8080/soap/HelloWS/5678</ResourceID>
    <TrustAnchor>http:// 192.168.11.2:8080/soap/</TrustAnchor>
  </ServiceDescription>

</SenConfiguration>
```

Example 9-1. Service descriptions in `SenSessions.xml`.

The configuration in Example 9-1 defines three services. The three services are an IDP, a discovery service, and an actual application service. In addition, it describes a framework (Liberty ID-WSF).[2]

Note, that access to the actual configuration files is restricted. You can only change these files by accessing the NWSF API functions defined for registering, associating, and inquiring services. Do not attempt to modify these files directly.

9.3 XML Processing in NWSF

Before discussing the actual topic of the chapter, the Java APIs, we examine the NWSF for Java XML processing capabilities. Since the underlying Java platform does not provide XML parsing capabilities for Java applications, NWSF for Java offers the `com.nokia.mws.wsf.xml` package to simplify the parsing operations. Two code examples in this chapter demonstrate the usage of these classes. Section 7.7 introduces XML processing and describes the relevant SOA for S60 platform APIs.

In general, Web service client applications explicitly use XML parsing in two basic scenarios: when attempting to deal with content received as an XML message, and when producing content to be sent as an XML message.

The classes and interfaces in the XML package are as follows:

1. `Element` – This interface represents an element in the XML document tree.
2. `Attribute` – This interface represents XML attributes.
3. `ElementFactory` – This class contains factory methods for instantiating objects that implement the Element interface.
4. `ElementException` – Exception resulting from errors in XML processing.

The XML handler implementation in the `com.nokia.mws.wsf.xml` package is not fully compliant with the W3C XML specification, because it is mainly optimized for protocol message handling. The implementation has the following limitations:

- The interface does not support the handling of mixed content XML (`Element` has only one content field). When parsing XML, the interspersed content is concatenated and returned as a single piece of content.
- When parsing XML documents, content that contains only white space characters is ignored.
- The interface does not support XML comments, processing instructions, or CDATA sections.

Note, that in order to use NWSF for Java to access Web services, it is also possible to use any other XML handling feature available on the device, or to download some of the publicly available XML parser implementations to assist XML handling.

9.3.1 Using Element to Create New XML Messages

The following example code creates a RequestHello XML message, which is later used in a basic Web service messaging example:

```
Element e = ElementFactory.getInstance()
    .createElement(null, "RequestHello", "urn:nokia:ws:samples");
e.addElement("HelloString").setContent(
    "Hello Web Service Provider!");
String helloRequest = e.toXml();
```

Example 9-2. NWSF for Java code used to create a RequestHello XML message.

The resulting XML document looks like this:

```
<RequestHello xmlns="urn:nokia:ws:samples">
  <HelloString>Hello Web Service Provider!</HelloString>
</RequestHello>
```

Example 9-3. The RequestHello XML message.

9.3.2 Using Element to Parse Incoming Messages

To parse an XML message string, create the following code:

```
String xmlString =
    "<wsse:Security xmlns:wsse=\"http://docs.oasis-open.org/wss/2004/01/
oasis-200401-wss-wssecurity-secext-1.0.xsd\">" +
    "    <wsse:UsernameToken>" +
    "        <wsse:Username>username</wsse:Username>" +
    "        <wsse:Password>password</wsse:Password>" +
    "    </wsse:UsernameToken>" +
    "</wsse:Security>";

Element e = ElementFactory.getInstance().parseElement(xmlString);
```

Example 9-4. Parsing an XML string.

Once you get the `Element` object, you have two alternative ways to access child elements in the tree: either access the child element directly by its name, or use the `GetNextElement` iterator to move through the entire tree. Please note that the name-based method does not provide instant access to an arbitrary element in the tree. To access the element contents, you first need to navigate to the element in question with a `getElement(...)` method (for information, see the JavaDocs).

Below is an example on how to access an individual child element:

```
// Get username and password from the security header.
String username = e.getElement("UsernameToken")
    .getElement("Username").getContent();
String password = e.getElement("UsernameToken")
    .getElement("Password").getContent();
System.out.println("Username=" + username +
                   "\nPassword=" + password);
```

Example 9-5. Accessing Username and Password elements.

Below is an example on how to move through the element tree using the iterator:

```
// Get username and password from the security header
// using getNextElement iterator.
while (e != null) {
    if (e.getName().equals("Username")) {
        System.out.println("Username=" + e.getContent());
    }
    else if (e.getName().equals("Password")) {
        System.out.println("Password=" + e.getContent());
    }
    // Get the next element.
    e = e.getNextElement();
}
```

Example 9-6. Iterating through an element tree.

CASE STUDY: COMMUNITY INFORMATION SHARE SERVICE

The Community Information Share Service (CISS) is a joint research project of Vodafone Group R&D, Trustgenix, and Nokia. The proof-of-concept demonstration shows how the user is able to manage her communities with the help of a mobile device, including the possibility to access information on the community members' devices (such as the calendar). For that to be possible, Web service provider functionality is required – the device needs to be able to respond to requests concerning calendar information. The CISS demonstration is a good example of implementations based on the Liberty ID-WSF specifications, featuring the Authentication Service, Discovery Service, Interaction Service, PAOS (reverse SOAP binding), and service invocation.

With the CISS client, users can:

- Create, update, and delete communities
- Register with a community
- View and search members
- Invite users to existing communities (from the address book on the device)
- Manage agendas (including invitations to meetings) and publish community events

The implementation on the device contains the Community Application, featuring the Community User Interface, the community manager, the agenda client, and the personal agenda manager. All of these running on top of the Nokia Web Services Framework, featuring support for both Web service client and provider functionality.

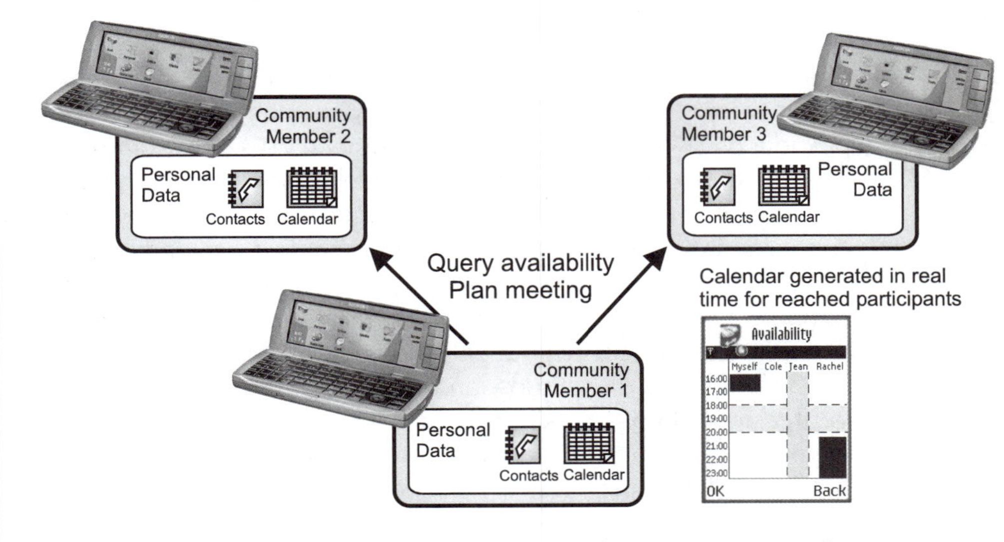

Figure 9-2. Providing calendar information from community members.

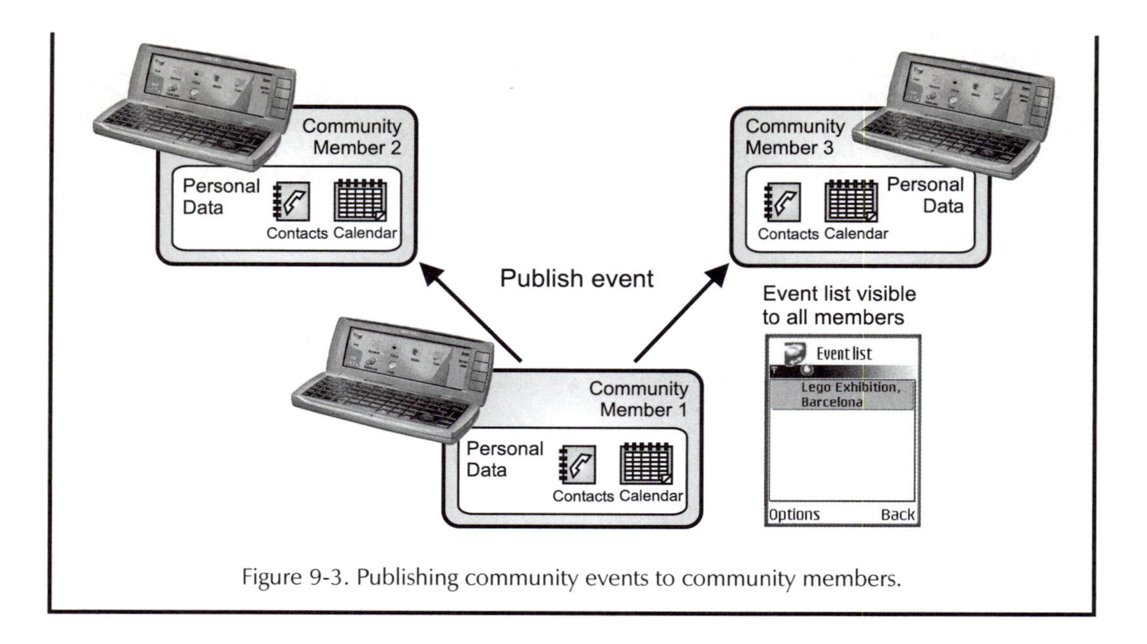

Figure 9-3. Publishing community events to community members.

9.4 HelloWS Example

The following example introduces the NWSF Java API. There are generally two entities involved in Web services interaction: a Web service provider (WSP) and a Web service consumer (WSC). The example shows how to construct a WSC for a simple "Hello World" Web service.

Note, that we use the same example to demonstrate all of the connection varieties, identity and security features included. In addition, this chapter provides a set of dummy server side codes to enable the testing of the client. Hence, the basic example below does not contain the complete client code, but only the `basicSample()` method. This method, advanced methods, and the Java Servlet code samples are located in section 9.8. All of the advanced code snippets are shown below before we introduce the complete code example.

The HelloWS messages are defined by the following XML schema document. Building an XML schema document is not required for describing a message exchange, but it usually helps in understanding the data sent in the messages.

```
<?xml version="1.0" encoding="UTF-8"?>
<xs:schema targetNamespace="urn:nokia:ws:samples"
  xmlns:xs="http://www.w3.org/2001/XMLSchema"
  xmlns="urn:nokia:ws:samples"
  version="1.0-01">

  <xs:complexType name="RequestHelloType">
```

```
    <xs:element name="HelloString" type="xs:string">
  </xs:complexType>
  <xs:element name="RequestHello" type="RequestHelloType"/>

  <xs:complexType name="HelloResponseType">
    <xs:element name="ResponseString" type="xs:string">
  </xs:complexType>

  <xs:element name="HelloResponse" type="HelloResponseType"/>

</xs:schema>
```

Example 9-7. XML Schema document describing HelloWS messages.

In this very simple example, the WSC sends a `<RequestHello>` message to the Web service. The WSP responds with a `<HelloResponse>` message. The message exchange is shown in Example 9-8 below.

1. NWSF WSC sends a Hello Request:

```
POST /soap/HelloWS HTTP/1.1
Host:
Accept: text/xml
Expect: 100-continue
User-Agent: Sen
Content-Length: 203
Content-Type: text/xml
SOAPAction:

<S:Envelope xmlns:S="http://schemas.xmlsoap.org/soap/envelope/">
  <S:Body>
    <RequestHello xmlns="urn:nokia:ws:samples">
      <HelloString>Hello Web Service Provider!</HelloString>
    </RequestHello>
  </S:Body>
</S:Envelope>
```

2. HelloWS Web service responds:

```
HTTP/1.1 100 Continue

HTTP/1.1 200 OK
Date: Thu, 23 Sep 2004 19:57:39 GMT
Server: Jetty/4.2.21 (Windows 2000/5.0 x86 java/1.4.2_04)

Content-Type: text/xml
Content-Length: 270
<S:Envelope xmlns:S="http://schemas.xmlsoap.org/soap/envelope/">
  <S:Body>
    <ex:HelloResponse xmlns:ex="urn:nokia:ws:samples">
      <ex:ResponseString>
        Hello Web Service Consumer!
      </ex:ResponseString>
    </ex:HelloResponse>
  </S:Body>

</S:Envelope>
```

Example 9-8. Message exchange related to HelloWS.

The message flow in Example 9-8 was generated with the following main client code (the complete code example with advanced features is presented later in section 9.8):

```
public void basicSample() {
    try {
        WSFObjectFactory objectFactory = WSFObjectFactory.getInstance();
        ServiceManager serviceManager = objectFactory.getServiceManager();

        ServiceDescription serviceDesc =
            objectFactory.getServiceDescription(
                "http:// 192.168.11.2:8080/mystuff/HelloWS",
                "urn:nokia:ws:samples",
                ServiceManager.BASIC_WS_FRAMEWORK_ID);
        ServiceConnection connection =
            serviceManager.getServiceConnection(serviceDesc, null);

        Element e = ElementFactory.getInstance()
            .createElement(null, "RequestHello", "urn:nokia:ws:samples");
        e.addElement("HelloString").setContent(
            "Hello Web Service Provider!");
        String helloRequest = e.toXml();

        // Alternatively request can be constructed directly as a string:
        //String helloRequest =
        // "<RequestHello xmlns=\"urn:nokia:ws:samples\">" +
        // "<HelloString>Hello Web Service Provider!</HelloString>" +
        // "</RequestHello>";

        System.out.println("Sending request " + helloRequest);
        String helloResponse = connection.submit(helloRequest);
        System.out.println("Response is " + helloResponse); }
    catch (ElementException ee) {
        System.out.println("Exception: " + ee.toString());
    }
    catch (WSFException wsfe) {
        System.out.println("Exception: " + wsfe.toString());
    }
}
```

Example 9-9. Main client code related to HelloWS.

In Example 9-9, the first boldfaced line creates a "service proxy" object. This constructor provides the code for connecting to a service. The second boldfaced line shows a literal text message being sent to the service.

Note the following issues:

1. Before a message is sent to the Web service provider, a connection to that provider's service has to be established and the service has to be described to the service provider.
2. NWSF uses the concept of a *framework* to send messages that are appropriate for a particular service. The NWSF framework concept is explained in section 7.2.1. In this simple example, the basic

framework can be used, because the service is quite straightforward. It does not require authentication or expect any additional SOAP headers.

3. The XML sent in the message is produced by creating `Element` objects that represent those pieces of XML (a miniature XML DOM tree). It is also possible to merely use strings, and submit XML strings directly to NWSF for processing and sending.

9.5 Connecting to Web Services

With the basic Web service client out of the way, we now take a closer look at the Web service messaging features provided by NWSF for Java. The following sections describe how to establish connections to Web services, by using and not using a framework.

Establishing a connection to a Web service with NWSF is simple, as demonstrated in the example above. Once the connection has been established, the application can simply submit messages to the service.

Note, that it is not possible to use one thread to establish a service connection and then use that connection from another thread. A service connection has to be used from the same thread that was used to establish it. Failure to observe this rule results in a kernel exception. Nevertheless, separating the service connection thread from the rest of the application is advisable as it ensures that, for example, the user interface remains responsive at all times.

The following line, taken from the HelloWS example, submits a message synchronously to the service: `String helloResponse = connection.submit(helloRequest);`

An object representing the service message is passed to NWSF for transmission to the service. If you do this using a framework, the SOAP envelope and any framework-specific headers are added by NWSF. It is, however, possible to submit a full SOAP envelope to NWSF. You should only do this if you intend to add all necessary headers yourself.

9.5.1 Connecting to Framework-Based Services

By selecting a framework to access a Web service, an application is indicating that the framework will add SOAP headers to a message. For example, the Liberty ID-WSF defines SOAP headers for message correlation (so that an application can determine that one message is linked to another through a `messageID` value). This correlation header is automatically added to messages produced using that framework.

Frameworks can be plugged into the Service Development API, which means that it is possible to add new frameworks to handle different ways of accessing a Web service. Currently, the only framework plugged into the Service Development API is one for the Liberty ID-WSF specifications. This framework is listed in the Service Development API `SenSessions.xml` configuration file:

```
<Framework xmlns="urn:com.nokia.Sen.idwsf.config.1.0" class="com.nokia.Sen.
idwsf.IdentityBasedWebServicesFramework"/>
```

Example 9-10. Configuring the ID-WSF framework plug-in for NWSF.

This framework implements the Liberty ID-WSF specifications, allowing a developer to simply ask for the framework to be used when connecting to a Liberty-compliant Web service. In this particular case, the connection may require the client application first to authenticate itself to the service possibly via a third party (through an associated IDP) and acquire credentials.

Frameworks can be specified with a framework identifier. For the Liberty framework, the identifier is `ServiceManager.LIBERTY_WS_FRAMEWORK_ID`. Note that if no framework identifier is specified in the service description used to create the service connection, the Liberty framework is automatically used to connect to the service, because this is the default framework.

Note, that the `com.nokia.mws.wsf.SOAPMessage` object cannot be submitted to service connections using the Liberty framework. Instead, requests have to be submitted as strings. For basic Web services, a request can be submitted either as a string or as `com.nokia.mws.wsf.SOAPMessage`.

9.5.2 Connecting to Web Services without Using a Framework

A basic Web service does not require any particular framework (i.e., special SOAP headers) or a well-defined policy to access a Web service – in fact, connecting to such services is quite simple. Nevertheless, since the default policy is to use a framework, the developer must specifically request NWSF not to use a framework. When creating connections to basic Web services and creating a `ServiceDescription`, use the following framework identifier: `ServiceManager.BASIC_WS_FRAMEWORK_ID`. The use of basic Web services requires application developers to add all the necessary SOAP headers (for authentication, addressing, etc.) to their messages before submitting those messages to NWSF.

9.6 Using the Service Database

The service database is where services can be stored persistently between invocations of the Service Development API framework. The Service Manager reads these descriptions when it starts up, and the descriptions can then be requested by client applications.

Services can be registered explicitly. This is very useful if a service that you are working on uses, for example, a particular discovery service. You can use the Service Development API to register both the actual service and the discovery service.

Registering a service is particularly useful, if a service requires authentication information to identify the user – in other words, credentials. Once the credentials have been acquired, and they are valid for a relatively long period, the Service Development API caches a session with a service, so that the information will be available for the next session. This improves performance, as it cuts down on the number of network requests that need to be submitted.

You can also register a service during the installation of a `.sis` package. One way to do this is to make a separate executable that performs registrations and associations, including the executable in the installation package, and then to run the executable using the `FR` (File Run) parameter.

Registering a service is a straightforward process, which includes the following two phases:

1. Establish a connection to the Service Manager.
2. Ask the Service Manager to register a particular service as follows:

```
WSFObjectFactory objectFactory = WSFObjectFactory.getInstance();
ServiceManager serviceManager = objectFactory.getServiceManager();

ServiceDescription serviceDesc =
    objectFactory.getServiceDescription(
        "http://192.168.11.2:8080/mystuff/HelloWS", null, null);
serviceManager.registerServiceDescription(serviceDesc);
```

Example 9-11. Registering a service with the Service Manager.

Once you have completed the registration, the framework can store service descriptions persistently.

Note, that service descriptions can only be registered for services that are accessed using the Liberty framework. Service descriptions for basic Web services do not have to be – and cannot be – registered because they do not contain any data that could be reused or shared between multiple connections.

9.7 Identity and Security

The identity concepts and Web service security technologies were introduced in chapter 5. This section builds on that knowledge, and connects the concepts to corresponding NWSF features. In this section, we describe how to register an IDP and associate it with another service. In addition, we provide several examples of different levels of authentication ranging from HTTP Basic Authentication to IDP-based scenarios.

9.7.1 Registering an Identity Provider

IDPs can be added to the framework using a `registerIdentityProvider()` API method call. In practice, the complete task involves three steps:

1. Obtain the Service Manager, which is used to realize the registration.
2. Create the actual IDP object.
3. Provide the Service Manager with the object by using the `registerIdentityProvider()` method call.

The code required to complete these steps appears as follows:

```
WSFObjectFactory objectFactory = WSFObjectFactory.getInstance();
ServiceManager serviceManager = objectFactory.getServiceManager();

IdentityProvider idP =
    objectFactory.getIdentityProvider(
        "https:// 192.168.11.2", // id
        "https:// 192.168.11.2:8181/soap/IdPAS", // endpoint
        "urn:liberty:as:2004-04", // contract
        null); // frameworkId, default is LIBERTY_WS_FRAMEWORK_ID
serviceManager.registerIdentityProvider(idP);
```

Example 9-12. Registering identity providers.

A client application, which used to connect to a specific service that uses this IDP, might register the provider when the application is installed.

9.7.2 Associating an Identity Provider with another Service

IDPs are generally used to provide authentication services to other service providers. A service provider probably only cooperates with certain IDPs. Therefore, a service operated by a certain service provider may only accept an authentication assertion issued by an IDP run by the same business entity, or a company with which it has a business agreement (a business partner, for example). In order to tell the framework that a particular service requires the use of a certain IDP (i.e., that the service provider trusts the IDP), this IDP must be associated with the service. The actual association means linking the IDP identifier with the service provider identifier.

The following code sample demonstrates how to link the IDP identifier with the service provider identifier (either a specific endpoint, or the contract URI) of the service being associated. We first register both identifiers with the `serviceManager`, and then link the identifiers to each other with an `associateService()` method call.

```
serviceManager.registerServiceDescription(serviceDesc);
serviceManager.registerIdentityProvider(idP);

serviceManager.associateService(serviceDesc, idP);
```

Example 9-13. Linking an IDP with an SP.

Above, the IDP identifier (`idP`) is associated with `serviceDesc`, either an actual service endpoint (which associates the IDP only with that single service endpoint) or a service contract (allowing one to associate the IDP with all services adhering to a certain schema contract, such as `urn:nokia:ws:samples`).

9.7.3 Using IDP-Based Authentication

In the HelloWS example, the `RequestHello` message was submitted without using any framework and without authentication to the service. However, the operator of the Web service may want to authenticate the users of the service by using an existing IDP. In this case, using authenticated access to the service is easy from perspective of both the client and the service provider. Note, however, that the IDP has to be registered to the framework, and that the service has to be associated with the IDP, as described above in sections 9.7.1 and 9.7.2.

It should be noted that transforming the HelloWS client application into an IDP-based version requires minimal changes. In fact, all that needs to be done is to change the service description constructor to read as follows:

```
ServiceDescription serviceDesc =
    objectFactory.getServiceDescription(
        null, // endpoint
        "urn:nokia:ws:samples", // contract
        null); // default is ServiceManager.LIBERTY_WS_FRAMEWORK_ID
```

Example 9-14. Service description constructor for IDP-based authentication.

Those simple code changes result in the following long exchange:

1. Get authenticated in order to locate a discovery service.

```
POST /soap/IDPAS HTTP/1.1
Host:
Accept: text/xml
Expect: 100-continue
User-Agent: Sen
Content-Length: 370
Content-Type: text/xml
SOAPACTION:

<S:Envelope xmlns:S="http://schemas.xmlsoap.org/soap/envelope/">

  <S:Header>
    <sb:Correlation xmlns:sb="urn:liberty:sb:2003-08"
                    messageID="C8797D0D-9020-07FC-AF0A-5622C01F4A61"
                    timestamp="2004-09-27T14:50:45Z"/>
  </S:Header>

  <S:Body>
    <sa:SASLRequest xmlns:sa="urn:liberty:sa:2004-04"
                    mechanism="PLAIN ANONYMOUS CRAM-MD5"
                   advisoryAuthnID="012345678901234"/>
  </S:Body>

</S:Envelope>
```

2. Response from the Authentication Service (AS), asking Service Development API to continue with SASL PLAIN authentication.

```
HTTP/1.1 100 Continue
HTTP/1.1 200 OK
Date: Mon, 27 Sep 2004 18:50:59 GMT
Server: Jetty/4.2.21 (Windows 2000/5.0 x86 java/1.4.2_04)
Content-Type: text/xml
Content-Length: 466

<S:Envelope xmlns:S="http://schemas.xmlsoap.org/soap/envelope/">
  <S:Header>

    <sb:Correlation s:mustUnderstand="1"
                    xmlns:sb="urn:liberty:sb:2003-08"
                    id="thisCorrHdr.2345"
                    messageID="i48b4353f50aca1494665d61b93498c885449c868"
                    refToMessageID="C8797D0D-9020-07FC-AF0A-5622C01F4A61"
                    timestamp="2004-02-03T22:12:27Z"/>
  </S:Header>

  <S:Body>

    <SASLResponse serverMechanism="PLAIN"
                  xmlns="urn:liberty:sa:2004-04">
      <Status code="continue"/>

    </SASLResponse>

  </S:Body>
</S:Envelope>
```

3. Continuing interaction with AS.

```
POST /soap/IDPAS HTTP/1.1
Host:
Accept: text/xml
Expect: 100-continue
User-Agent: Sen
Content-Length: 455
Content-Type: text/xml
SOAPAction:

<S:Envelope xmlns:S="http://schemas.xmlsoap.org/soap/envelope/">
  <S:Header>
    <sb:Correlation xmlns:sb="urn:liberty:sb:2003-08"
                    messageID="D93D6E95-A1F0-4A50-7938-1D1FA9D77918"
refToMessageID="i48b4353f50aca1494665d61b93498c885449c868"
                    timestamp="2004-09-27T14:51:00Z"/>
  </S:Header>

  <S:Body>
    <sa:SASLRequest xmlns:sa="urn:liberty:sa:2004-04"
                    mechanism="PLAIN">
      <sa:Data>ADAxMjM0NTY3ODkwMTIzNAAwMTIzNDU2Nzg5MDEyMzQ=</sa:Data>
    </sa:SASLRequest>
  </S:Body>

</S:Envelope>
```

4. AS responds, accepting the authentication, and providing a resource offering for the discovery service (DS).

```
HTTP/1.1 100 Continue
HTTP/1.1 200 OK
Date: Mon, 27 Sep 2004 19:01:51 GMT
Server: Jetty/4.2.21 (Windows 2000/5.0 x86 java/1.4.2_04)
Content-Type: text/xml
Content-Length: 1169

<S:Envelope xmlns:S="http://schemas.xmlsoap.org/soap/envelope/">
  <S:Header>
    <sb:Correlation s:mustUnderstand="1"
                    xmlns:sb="urn:liberty:sb:2003-08"
                    id="thisCorrHdr.3456"
                    messageID="e44b8753f05abb1499657e61b83378c775219a768"
                    refToMessageID="D93D6E95-A1F0-4A50-7938-1D1FA9D77918"
                    timestamp="2004-02-03T22:12:27Z" />
  </S:Header>
  <S:Body>
    <SASLResponse serverMechanism="PLAIN"
                  xmlns="urn:liberty:sa:2004-04">
      <Status code="sa:OK"/>
      <ResourceOffering entryID="1">

<ResourceID>http://90.0.0.18:8080/soap/012345678901234</ResourceID>
       <ServiceInstance>
       <ServiceType>urn:liberty:disco:2003-08</ServiceType>
       <ProviderID>http://90.0.0.18:8080/soap/</ProviderID>
       <Description>
         <CredentialRef>2sxJu9g/vvLG9sAN9bKp/8q0NKU=</CredentialRef>
         <SecurityMechID>urn:liberty:security:2004-04:null:Bearer</
SecurityMechID>
         <Endpoint>http://90.0.0.18:8080/soap/IDPDS</Endpoint>
       </Description></ServiceInstance>
          </ResourceOffering>
```

```
         <Credentials notOnOrAfter="2004-09-28T18:28:44Z">
           <saml:Assertion xmlns:saml="urn:oasis:names:tc:SAML:1.0:
assertion"
                          AssertionID="2sxJu9g/vvLG9sAN9bKp/8q0NKU=">
           ...
           </saml:Assertion>
        </Credentials>
      </SASLResponse>
   </S:Body>
</S:Envelope>
```

5. Service Development API (NWSF) asks DS whether it knows the service type "urn: nokia:ws:samples".

```
POST /soap/IDPDS HTTP/1.1
Host:
Accept: text/xml
Expect: 100-continue
User-Agent: Sen
Content-Length: 679
Content-Type: text/xml
SOAPAction:

<S:Envelope xmlns:S="http://schemas.xmlsoap.org/soap/envelope/">
  <S:Header>

    <sb:Correlation xmlns:sb="urn:liberty:sb:2003-08"
                    messageID="F5683B9F-DF73-AA9E-D01A-82CFFE8F6341"
                    timestamp="2004-09-27T14:51:01Z"/>
    <wsse:Security
xmlns:wsse="http://schemas.xmlsoap.org/ws/2003/06/secext">
      <saml:Assertion xmlns:saml="urn:oasis:names:tc:SAML:1.0:assertion"
                      AssertionID="2sxJu9g/vvLG9sAN9bKp/8q0NKU=">
        ...
      </saml:Assertion>
    </wsse:Security>

  </S:Header>

  <S:Body>

    <Query xmlns="urn:liberty:disco:2003-08">
      <ResourceID>http://90.0.0.18:8080/soap/012345678901234</ResourceID>
      <RequestedServiceType>
        <ServiceType>urn:nokia:ws:samples</ServiceType>
      </RequestedServiceType>
    </Query>

  </S:Body>
</S:Envelope>
```

6. DS knows the service type, and provides the information and credentials to allow access to an instance of that service.

```
HTTP/1.1 100 Continue
HTTP/1.1 200 OK
Date: Mon, 27 Sep 2004 19:01:53 GMT
Server: Jetty/4.2.21 (Windows 2000/5.0 x86 java/1.4.2_04)
Content-Type: text/xml
Content-Length: 1066

<S:Envelope xmlns:S="http://schemas.xmlsoap.org/soap/envelope/">
  <S:Body>
    <QueryResponse xmlns="urn:liberty:disco:2003-08">
```

```
      <Status code="ok"/>

      <ResourceOffering entryID="1">

<ResourceID>http://90.0.0.18:8080/soap/HelloWS/5678901234</ResourceID>
        <ServiceInstance>
          <ServiceType>urn:nokia:ws:samples</ServiceType>
          <ProviderID>http://90.0.0.18:8080/soap/</ProviderID>
          <Description>
            <CredentialRef>2sxJu9g/vvLG9sAN9bKp/8q0NKU=</CredentialRef>
            <SecurityMechID>urn:liberty:security:2004-04:null:Bearer</
SecurityMechID>
            <Endpoint>http://90.0.0.18:8080/soap/HelloWS</Endpoint>
          </Description>
        </ServiceInstance>
       </ResourceOffering>

       <Credentials notOnOrAfter="2004-09-28T18:28:44Z">
         <wsse:BinarySecurityToken EncodingType="wsse:Base64Binary"

wsu:Id="ia1575535f5b0712dbff7033db0721e4f838390f3"

xmlns:wsse="http://schemas.xmlsoap.org/ws/2003/06/secext"

xmlns:wsu="http://schemas.xmlsoap.org/ws/2003/06/utility">
          AZoOuAM4BdMxtKugmt1qiwZze11vQb/m5udOPOTa8Y5L
        </wsse:BinarySecurityToken>
      </Credentials>

    </QueryResponse>
  </S:Body>
</S:Envelope>
```

7. Service Development API (NWSF) WSC sends a Hello Request.

```
POST /soap/HelloWS HTTP/1.1
Host:
Accept: text/xml
Expect: 100-continue
User-Agent: Sen
Content-Length: 528
Content-Type: text/xml
SOAPAction:

<S:Envelope xmlns:S="http://schemas.xmlsoap.org/soap/envelope/">
  <S:Header>
    <wsse:Security xmlns:wsse="http://schemas.xmlsoap.org/ws/2003/06/
secext">
      <wsse:BinarySecurityToken EncodingType="wsse:Base64Binary"

wsu:Id="ia1575535f5b0712dbff7033db0721e4f838390f3"

xmlns:wsse="http://schemas.xmlsoap.org/ws/2003/06/secext"

xmlns:wsu="http://schemas.xmlsoap.org/ws/2003/06/utility">
        AZoOuAM4BdMxtKugmt1qiwZze11vQb/m5udOPOTa8Y5L
      </wsse:BinarySecurityToken>
    </wsse:Security>
  </S:Header>
  <S:Body>
    <RequestHello xmlns="urn:nokia:ws:samples">
      <HelloString>Hello Web Service Provider!</HelloString>
    </RequestHello>
  </S:Body>
</S:Envelope>
```

8. HelloWS Web Service responds.

```
HTTP/1.1 100 Continue
HTTP/1.1 200 OK
Date: Thu, 23 Sep 2004 19:02:39 GMT
Server: Jetty/4.2.21 (Windows 2000/5.0 x86 java/1.4.2_04)
Content-Type: text/xml
Content-Length: 270

<S:Envelope xmlns:S="http://schemas.xmlsoap.org/soap/envelope/">
  <S:Body>
    <ex:HelloResponse xmlns:ex="urn:nokia:ws:samples">
      <ex:ResponseString>
        Hello Web Service Consumer!
      </ex:ResponseString>
    </ex:HelloResponse>
  </S:Body>
</S:Envelope>
```

Example 9-15. Message exchange in IDP-based authentication.

This eight-message exchange can be broken into four separate groups of two messages. In the first two pairs of exchange, the client application interacts with the IDP concerning the requested service. The IDP offers a Liberty Alliance Authentication Service interface, which the client uses to authenticate itself.

The message exchange proceeds as follows:

1. The client tells the IDP that it is requesting SASL authentication, and can perform the SASL mechanisms PLAIN, ANONYMOUS, and CRAM-MD5.
2. The service responds by requesting the client to perform the PLAIN mechanism.
3. The client agrees, returning the base-64-encoded credentials to the service.
4. Once the authentication service has authenticated the requester, it tells the requester where to go to discover services available to the user. It returns a description of the user's discovery service: an identity-based Web service into which other Web services can register information about themselves.
5. The Service Development API framework sends a request to the discovery service, enquiring about services of the type specified by the client application (`urn:nokia:ws:samples`).
6. The discovery service responds, offering a credential required for access to the requested service, security policy information for the service, and information regarding the service endpoint.

7. So far, all of the work has been carried out by the Service Development API on behalf of the client application, without any interaction by the client application itself. However, at this stage the actual client message is sent to the service.
8. A response is received, and it can be processed by the service consumer application.

9.7.4 Authenticating SOAP Messages with WS-Security Username Token

The Service Development API contains an implementation of the WS-Security specification, which allows for the authentication of SOAP messages using a security token. One type of security token is the `UsernameToken`, which is used simply to provide a username and a password in order to authenticate the SOAP message. Note that this is not the most secure authentication method available, and when sending credentials like these, SSL/TLS should be used to protect the content of the message. This type of authentication is similar to HTTP Basic Authentication, but is done at the SOAP rather than the HTTP layer.

The following code excerpt shows how the security token is added to a SOAP message:

```
Element secHeader = ElementFactory.getInstance()
    .createElement("wsse", "Security",
        "http://docs.oasis-open.org/wss/2004/01/oasis-200401-wss-wssecurity-
secext-1.0.xsd");
Element usernameToken = secHeader.addElement(null, "UsernameToken",
        "http://docs.oasis-open.org/wss/2004/01/oasis-200401-wss-wssecurity-
secext-1.0.xsd");
usernameToken.addElement("Username").setContent("username");

Element body = ElementFactory.getInstance()
    .createElement(null, "RequestHello", "urn:nokia:ws:samples");
body.addElement("HelloString").setContent("Hello Web Service Provider!");

SOAPMessage message = new SOAPMessage();
message.setSecurityToken(secHeader.toXml());
message.setBody(body.toXml());
```

Example 9-16. Adding a security token to a SOAP message.

The security token has now been added to the SOAP message. The following code shows what happens when Service Development API sends the message:

```
POST /soap/HelloWS HTTP/1.1
Host:
Accept: text/xml
Expect: 100-continue
User-Agent: Sen
```

```
Content-Length: 426
Content-Type: text/xml
SOAPAction:

<S:Envelope xmlns:S="http://schemas.xmlsoap.org/soap/envelope/">
  <S:Header>
    <wsse:Security xmlns:wsse="http://docs.oasis-open.org/wss/2004/01/oasis-
200401-wss-wssecurity-secext-1.0.xsd">
      <UsernameToken xmlns="http://docs.oasis-open.org/wss/2004/01/oasis-
200401-wss-wssecurity-secext-1.0.xsd">
        <Username>username</Username>
      </UsernameToken>
    </wsse:Security>
  </S:Header>
  <S:Body>
    <RequestHello xmlns="urn:nokia:samples">
      <HelloString>Hello Web Service Provider!</HelloString>
    </RequestHello>
  </S:Body>
</S:Envelope>
```

Example 9-17. Service Development API sends the SOAP message with a security token.

Note that this method is an alternative to the previous Liberty-based example; this method uses the basic framework.

9.7.5 Self-Authentication with HTTP Basic Authentication

Some Web services do not use third-party authentication via a framework, but still want to authenticate the requester. One way of accomplishing this is called HTTP Basic Authentication. The Service Development API supports the relaying of usernames and passwords using HTTP Basic Authentication. In practice, NWSF utilizes user authentication information to instantiate an IDP of its own. This NWSF-hosted IDP provides the required credentials to the service provider. When using this scheme, service descriptions do not need to be associated with the IDP.

The following code excerpt shows how to authenticate the requester. After initializing the connection parameters and the `serviceManager`, we instantiate an IDP. The `getIdentityProvider()` method call takes the connection parameters and credentials as parameters.

```
String serviceEndpoint = "http://192.168.11.2:8080/mystuff/HelloWS";
String serviceContract = "urn:nokia:ws:samples";
String serviceId = "https:// 192.168.11.2";

WSFObjectFactory objectFactory = WSFObjectFactory.getInstance();
ServiceManager serviceManager = objectFactory.getServiceManager();

IdentityProvider idP =
     objectFactory.getIdentityProvider(
         serviceId,
         serviceEndpoint,
         serviceContract,
         ServiceManager.BASIC_WS_FRAMEWORK_ID);
```

```
idP.setUserInfo("username", "username", "password");
serviceManager.registerIdentityProvider(idP);

ServiceDescription serviceDesc =
    objectFactory.getServiceDescription(
        serviceEndpoint,
        serviceContract,
        ServiceManager.BASIC_WS_FRAMEWORK_ID);

ServiceConnection connection =
    serviceManager.getServiceConnection(serviceDesc, null);
```

Example 9-18. Authenticating the requester by using HTTP Basic Authentication.

In this example, the Web service endpoint is protected with HTTP Basic Authentication. When the initial request is made, the service provider returns an HTTP 401 error to the client, and requests authentication. The NWSF then instantiates an IDP service, and supplies the service provider with the credentials as described above.

The following example describes the message exchange of a service not only protected with HTTP Basic Authentication, but also with SOAP message security (in the form of a WS-Security `UsernameToken`).

1. *Initial request, posting the SOAP message.*

```
POST /protected/HelloWS HTTP/1.1
Host: 90.0.0.18:8080
Accept: text/xml
Expect: 100-continue
User-Agent: Sen
Content-Length: 623
Content-Type: text/xml
SOAPACTION: ""

<S:Envelope xmlns:S="http://schemas.xmlsoap.org/soap/envelope/">

  <S:Header>
    <wsse:Security xmlns:wsse="http://docs.oasis-open.org/wss/2004/01/oasis-
200401-wss-wssecurity-secext-1.0.xsd">
      <UsernameToken xmlns="http://docs.oasis-open.org/wss/2004/01/oasis-
200401-wss-wssecurity-secext-1.0.xsd">
        <Username>username</Username>
        <Password>password</Password>
      </UsernameToken>
    </wsse:Security>

  </S:Header>

  <S:Body>
    <RequestHello xmlns="urn:nokia:samples">
      <HelloString>Hello Web Service Provider!</HelloString>
    </RequestHello>
  </S:Body>
</S:Envelope>
```

2. The server responds, stating that the request was unauthorized, and prompting for credentials.

```
HTTP/1.1 401 Unauthorized
WWW-Authenticate: Basic realm="90.0.0.18:8080"
Content-Type: text/xml
Date: Thu, 30 Sep 2004 16:59:15 GMT
Server: Jetty/4.2.21 (Windows 2000/5.0 x86 java/1.4.2_04)

<html>
  <body>Unauthorized!</body>
</html>
```

3. NWSF adds the HTTP basic credentials to the request and re-sends.

```
POST /protected/HelloWS HTTP/1.1
Host: 90.0.0.18:8080
Accept: text/xml
Authorization: Basic dXNlcm5hbWU6cGFzc3dvcmQ=
Expect: 100-continue
User-Agent: Sen
Content-Length: 623
Content-Type: text/xml
SOAPACTION:

<S:Envelope xmlns:S="http://schemas.xmlsoap.org/soap/envelope/">

  <S:Header>

    <wsse:Security xmlns:wsse="http://docs.oasis-open.org/wss/2004/01/oasis-
200401-wss-wssecurity-secext-1.0.xsd">
      <UsernameToken xmlns="http://docs.oasis-open.org/wss/2004/01/oasis-
200401-wss-wssecurity-secext-1.0.xsd">
        <Username>username</Username>
        <Password>password</Password>
      </UsernameToken>
    </wsse:Security>

  </S:Header>

  <S:Body>
    <RequestHello xmlns="urn:nokia:samples">
      <HelloString>Hello Web Service Provider!</HelloString>
    </RequestHello>
  </S:Body>

</S:Envelope>
```

4. The server authenticates the request and responds affirmatively.

```
HTTP/1.1 100 Continue
HTTP/1.1 200 OK
Date: Thu, 30 Sep 2004 16:59:17 GMT
Server: Jetty/4.2.21 (Windows 2000/5.0 x86 java/1.4.2_04)
Content-Type: text/xml
Content-Length: 280

<S:Envelope xmlns:S="http://schemas.xmlsoap.org/soap/envelope/">
  <S:Body>
    <ex:HelloResponse xmlns:ex="urn:nokia:ws:samples">
      <ex:ResponseString>
        Hello Web Service Consumer!
      </ex:ResponseString>
    </ex:HelloResponse>
  </S:Body>
</S:Envelope>
```

Example 9-19. Message exchange for authentication request.

9.7.6 Policy and Services

As explained in chapter 4, Web services have different policies regarding the way they authenticate a requester. The policy can vary from requiring no authentication at all to requiring authentication both for the connection from the client (peer entity), and for the actual message(s) sent by the client. However, a service may also maintain other types of policies. In many cases – and particularly when using a framework – the implementation of the policy is handled by the framework.

In the case of the Liberty ID-WSF framework plug-in, the policy regarding the use of various security mechanisms (i.e., which peer-entity and message authentication mechanisms clients should use) is stored by a discovery service, and the NWSF obtains the policies from the discovery service and stores them for use by client applications.

Client applications can, however, set specific policies for service providers. For instance, an application can specify which identity providers (IDPs) they trust to authenticate a service requestor. This can be accomplished by using the `ServiceManager.registerIdentityProvider(...)` method. The list of IDP identifiers (`trustAnchors`) that are associated with the requested service can be obtained by calling the `ServiceManager.getServiceConnection(...)` method.

9.8 HelloWS.java

The following HelloWS.java example code demonstrates the usage of four different kinds of services. We begin with the already familiar basic HelloWS example, and then proceed to advanced, more secure examples. Each service type example is placed in a method of its own, and each method is called in the `main()` function in the following order:

1. `basicSample` demonstrates how to use a basic Web service lacking all authentication.
2. `httpBasicSample` demonstrates how to use a basic Web service with HTTP Basic Authentication.
3. `registrationSample` demonstrates how to register and associate an IDP and a service description for the Liberty Web service example described below (and executed last).
4. `libertySample` demonstrates how to use a Liberty-compliant Web service. The example assumes that an IDP and a service description have been registered (as in `registrationSample`) before a connection to the Liberty service is established.

The HelloWS.java example application does not contain a UI. These example functions print out the response message with a `System.out.println` function, and the realization of the output depends on the development and test environment. For example, when using Eclipse and the Nokia 9500 emulator, the response is displayed on the Eclipse Console. You can implement more advanced and practical functions simply by replacing the aforementioned `System.out.println` lines with actual application logic code. Please note that in the HelloWS.java code below, you need to edit the endpoints according to the IP addresses and Servlet paths of your test network.

```
HelloWS.java
/*
 * Copyright (c) 2004 Nokia Corporation.
 */
import com.nokia.mws.wsf.IdentityProvider;
import com.nokia.mws.wsf.ServiceDescription;
import com.nokia.mws.wsf.ServiceConnection;
import com.nokia.mws.wsf.ServiceManager;
import com.nokia.mws.wsf.WSFException;
import com.nokia.mws.wsf.WSFObjectFactory;
import com.nokia.mws.wsf.xml.Element;
import com.nokia.mws.wsf.xml.ElementException;
import com.nokia.mws.wsf.xml.ElementFactory;

/**
 * Sample code demonstrating the use of Java Web Services Framework API.
 */
public class HelloWS {
    private WSFObjectFactory objectFactory;
    private ServiceManager serviceManager;
    /**
     * Constructor.
     *
     * @throws WSFException if the ServiceManager cannot be
     * constructed or initialized
     */
    public HelloWS() throws WSFException {
        objectFactory = WSFObjectFactory.getInstance();
        serviceManager = objectFactory.getServiceManager();
    }

    /**
     * Main function.
     *
     * @param args
     */
    public static void main(String[] args) {
        try {
            HelloWS hws = new HelloWS();
            hws.basicSample();
            hws.httpBasicSample();
            hws.registrationSample();
            hws.libertySample();
        }
        catch (WSFException wsfe) {
           // Log error reason using wsfe.toString().
           System.out.println("Exception: " + wsfe.toString());
        }
    }
```

```
    /**
     * Sample code for basic Web service request.
     */
    public void basicSample() {
        try {
            // Get service description.
            ServiceDescription serviceDesc =
                objectFactory.getServiceDescription(
                    "http://192.168.11.2:8080/mystuff/HelloWS", // endpoint
                    "urn:nokia:ws:samples", // contract
                    ServiceManager.BASIC_WS_FRAMEWORK_ID); // frameworkId

            // Get connection to the service.
            ServiceConnection connection =
                serviceManager.getServiceConnection(serviceDesc, null);

            // Send the request, parse the response, and print it out to
the console.
            String response = sendHelloRequest(connection);
                 System.out.println(response);
            }
            catch (ElementException ee) {
                System.out.println("Exception: " + ee.toString());
            }
            catch (WSFException wsfe) {
                System.out.println("Exception: " + wsfe.toString());
            }
        }
        /**
         * Sample code for Liberty Web service request. Note that
         * registrationSample has to be executed once before running
         * libertySample.
         */
        public void libertySample() {
            try {
                // Get service description.
                ServiceDescription serviceDesc =
                    objectFactory.getServiceDescription(
                        null, // endpoint
                        "urn:nokia:ws:samples", // contract
                        null); // default is LIBERTY_WS_FRAMEWORK_ID
                // Get connection to the service.
                ServiceConnection connection =
                    serviceManager.getServiceConnection(serviceDesc, null);

               // Send the request, parse the response, and print it out to
the console.
               String response = sendHelloRequest(connection);
                    System.out.println(response);

            }
            catch (ElementException ee) {
               System.out.println("Exception: " + ee.toString());
            }
            catch (WSFException wsfe) {
               System.out.println("Exception: " + wsfe.toString());
            }
        }
        /**
         * Run the sample code for basic Web service request with
         * HTTP authentication.
         */
        public void httpBasicSample() {
            try {
               String serviceEndpoint =
                  "http://192.168.11.2:8080/mystuff/HelloWS";
```

```
                String serviceContract = "urn:nokia:ws:samples";
                String serviceId = "https://192.168.11.2";

                // Get identity provider (IDP) and register it for
                // HTTP basic authentication. Note that it is
                // enough to register an IDP once for each service.
                IdentityProvider idP =
                    objectFactory.getIdentityProvider(
                        serviceId,
                        serviceEndpoint,
                        serviceContract,
                        ServiceManager.BASIC_WS_FRAMEWORK_ID);
                idP.setUserInfo("username", "username", "password");
                serviceManager.registerIdentityProvider(idP);

                // Get service description.
                ServiceDescription serviceDesc =
                    objectFactory.getServiceDescription(
                        serviceEndpoint,
                        serviceContract,
                        ServiceManager.BASIC_WS_FRAMEWORK_ID);

                // Get connection to the service.
                ServiceConnection connection =
                    serviceManager.getServiceConnection(serviceDesc, null);

                // Send the request, parse the response, and print it out to
the console.
                String response = sendHelloRequest(connection);
                    System.out.println(response);
           }
           catch (ElementException ee) {
              System.out.println("Exception: " + ee.toString());
           }
           catch (WSFException wsfe) {
              System.out.println("Exception: " + wsfe.toString());
           }
       }
       /**
        * Sample code for registering IDP and service
        * descriptions.
        */
       public void registrationSample() {
           try {
               // Create and register IDP.
               IdentityProvider idP =
                   objectFactory.getIdentityProvider(
                       "https://192.168.11.2", // id
                       "https://192.168.11.2:8181/soap/IdPAS", // endpoint
                       "urn:liberty:as:2004-04", // contract
                       null); // default is LIBERTY_WS_FRAMEWORK_ID
               serviceManager.registerIdentityProvider(idP);

               // Create and register service description.
               ServiceDescription serviceDesc =
                   objectFactory.getServiceDescription(
                       "http:// 192.168.11.2:8080/mystuff/HelloWS", //
endpoint
                       "urn:liberty:as:2004-04", // contract
                       null); // default is LIBERTY_WS_FRAMEWORK_ID
               serviceManager.registerServiceDescription(serviceDesc);

               // Associate service to identity provider.
               serviceManager.associateService(serviceDesc, idP);
           }
           catch (WSFException wsfe) {
```

```
            System.out.println("Exception: " + wsfe.toString());
        }
    }
    /**
     * Sends Hello request to given connection and parses the response.
     *
     * @param connection service connection to which the request is sent.
     * @return response string from the Hello response
     * @throws ElementException if Element handling fails
     * @throws WSFException if submitting the request fails
     */
    public String sendHelloRequest(ServiceConnection connection)
            throws ElementException, WSFException {
        // Create request.
        Element e = ElementFactory.getInstance().createElement(
            null, "RequestHello", "urn:nokia:ws:samples");
        e.addElement("HelloString").setContent(
            "Hello Web Service Provider!");
        String helloRequest = e.toXml();

        // Alternatively request can be constructed directly as
        // a string:
        //String helloRequest =
        // "<RequestHello xmlns=\"urn:nokia:ws:samples\">"
        // + "<HelloString>Hello Web Service Provider!</HelloString>"
        // + "</RequestHello>";

        // Optimize performance by notifying framework that
        // a transaction is starting.
        connection.transactionStarting();

        // Submit the request.
        String helloResponse = connection.submit(helloRequest);

        // Notify framework that the transaction has been completed.
        connection.transactionCompleted();

        // Parse the response message.
        Element re = ElementFactory.getInstance()
            .parseElement(helloResponse);
        String responseContent = null;

        while(re!=null){
        if(re.getName().equals("ResponseString")){
         responseContent = re.getContent();
     }

     re = re.getNextElement();

     }
     return responseContent;
  }
}
```

Example 9-20. HelloWS.java code listing.

9.8.1 HelloWS.java Test Service Servlets

The HelloWS client code uses Java Servlets to allow the testing of the example code presented in this chapter. None of the following examples should be considered anything else than instructional examples: they do not represent any specific guidance on how to write such services, and should mostly be thought of as instructions on how *not to write* Web services. However, the examples do show how simple it is to write a Web service – it does not even require XML parsing.

These simple services could be quickly improved by, for example, parsing incoming messages (either as strings or as XML) and by generating message identifiers (perhaps using a GUID generator), which could be used by a recipient to index messages and prevent things like replay attacks.

Let us now examine the actual Java Servlet code snippets. We begin with a basic, non-authenticated connection, then present a dummy discovery service, and conclude with a dummy authentication service.

Please note that you need to edit the endpoints according to the actual test network IP addresses and Servlet paths. In addition, in a realistic environment that checks the lifetime of the credentials, the `notOnOrAfter` attribute values used in some Servlets have to be edited to enable the use of credentials.

The HelloWS Web service always returns a HelloResponse message, regardless of what message you send, or whether that service is authenticated. The `doPost()` method is shown below:

```
public void doPost(HttpServletRequest request, HttpServletResponse response)
      throws ServletException, IOException
    {
      PrintWriter out = response.getWriter() ;

      response.setContentType("text/xml") ;
      String outSoap = new String( "<S:Envelope xmlns:S=\"http://schemas.
xmlsoap.org/soap/envelope/\">\n" +
      " <S:Body>\n" +
      " <ex:HelloResponse xmlns:ex=\"urn:nokia:ws:samples\">\n" +
      " <ex:ResponseString>\n" +
      " Hello Web Service Consumer!\n" +
      " </ex:ResponseString>\n" +
      " </ex:HelloResponse>\n" +
      " </S:Body>\n" +
      "</S:Envelope>\n") ;

      response.setContentLength( outSoap.length() + 2 ) ;
      out.println( outSoap ) ;
}
```

Example 9-21. HelloResponse Java Servlet code listing.

Apart from the SOAP message output, the actual Servlet code for the dummy discovery service (called IDPDS in the example scenario) below is identical to that of the HelloWS Servlet. This Web service returns a `ResourceOffering` that describes the HelloWS service and provides a credential for accessing that service. In this service message, the SOAP header should contain a Liberty SOAP correlation header block, which references the message that prompted the response.

```
      String outSoap = new String( "<S:Envelope xmlns:S=\"http://schemas.
xmlsoap.org/soap/envelope/\">\n" +
      "  <S:Body>\n" +
      "    <QueryResponse xmlns=\"urn:liberty:disco:2003-08\">\n" +
      "      <Status code=\"ok\"/>" +
      "        <ResourceOffering entryID=\"1\">\n" +
      "          <ResourceID>http://192.168.11.2:8080/soap/
HelloWS/5678901234" +

      "</ResourceID>\n" +
      "<ServiceInstance>\n" +
      "<ServiceType>urn:nokia:ws:samples</ServiceType>\n" +
      "<ProviderID>http:// 192.168.11.2:8080/mystuff/</ProviderID>\n" +
      "<Description>\n" +
  "<CredentialRef>ia1575535f5b0712dbff7033db0721e4f838390f3</CredentialRef>"
+
      "<SecurityMechID>urn:liberty:security:2004-04:null:Bearer</
SecurityMechID>" +
      "<Endpoint>http:// 192.168.11.2:8080/mystuff/HelloWS</Endpoint>" +
      "</Description>" +
      "</ServiceInstance>" +
      "</ResourceOffering>" +
      "<Credentials notOnOrAfter=\"2004-09-28T18:28:44Z\">" +
      "<wsse:BinarySecurityToken EncodingType=\"wsse:Base64Binary\"" +
        " wsu:Id=\"ia1575535f5b0712dbff7033db0721e4f838390f3\"" +
        " xmlns:wsse=\"http://schemas.xmlsoap.org/ws/2003/06/secext\"" +
        " xmlns:wsu=\"http://schemas.xmlsoap.org/ws/2003/06/utility\">" +
        "AZoOuAM4BdMxtKugmt1qiwZze1lvQb/m5udOPOTa8Y5L" +
        "</wsse:BinarySecurityToken>\n" +
      "</Credentials>" +
      "</QueryResponse>" +
      "</S:Body>" +
      "</S:Envelope>\n") ;
```

Example 9-22. Dummy discovery service for `libertySample()`.

This simple dummy authentication service (called IdPAS in the example scenario) pretends to authenticate the requester using the SASL PLAIN mechanism. It is not very sophisticated, but it returns a credential and resource details that allow the requester to query an associated discovery service (IDPDS, as described above) about services of the type it is looking for (in this case, the HelloWS example).

Again, this example could be easily improved by adding means to generate unique message identifiers, and features to check for such an identifier in the incoming message. It could also be improved by having the service actually perform a verification of the authentication.

WARNING
Do not offer this particular authentication service in any real environment!

The `doPost()` method is illustrated in the following example:

```
public void doPost(HttpServletRequest request, HttpServletResponse response)
      throws ServletException, IOException
   {
      PrintWriter out = response.getWriter() ;
```

```
      response.setContentType("text/xml") ;

      String outSoap = new String(
     "<S:Envelope
xmlns:S=\"http://schemas.xmlsoap.org/soap/envelope/\">" +
     "<S:Header>" +
     "<sb:Correlation s:mustUnderstand=\"1\" xmlns:sb=\"urn:liberty:sb:2003-
08\"" +
     " id=\"thisCorrHdr.2345\"" +
     " messageID=\"i48b4353f50aca1494665d61b93498c885449c868\"" +
     " refToMessageID=\"-1909129211480725265-1455516650932757068\"" +
     " timestamp=\"2004-02-03T22:12:27Z\" />" +
     "</S:Header>" +
     "<S:Body>" +
     "<SASLResponse serverMechanism=\"PLAIN\"" +
     " xmlns=\"urn:liberty:sa:2004-04\">" +
     "<Status code=\"continue\"/>" +
     "</SASLResponse>" +
     "</S:Body>" +
     "</S:Envelope>") ;

     if (0 < request.getContentLength())
     {
     String line = null;

     // Translate all incoming bytes to characters.
     BufferedReader in = new BufferedReader
  ( new InputStreamReader (request.getInputStream()));

     // Read each line of input and add it to the output file.
     HttpUtils httpUtils = new HttpUtils();

     try
     {
        while (null != (line = in.readLine()))
        {
           try
           {
              if (line.indexOf( "sa:Data") > -1)
              {
                 outSoap = new String("<S:Envelope xmlns:S=\"http://schemas.
xmlsoap.org/soap/envelope/\">" +
                 "<S:Header>" +
                 "<sb:Correlation s:mustUnderstand=\"1\" xmlns:sb=\"urn:
liberty:sb:2003-08\"" +
                 " id=\"thisCorrHdr.2345\"" +
                 " messageID=\"i48b4353f50aca1494665d61b93498c885449c868\""
+
                 " refToMessageID=\"-1909129211480725265-
1455516650932757068\"" +
                 " timestamp=\"2004-02-03T22:12:27Z\" />" +
                 "</S:Header>" +
                 "<S:Body>" +
                 "<SASLResponse serverMechanism=\"PLAIN\"" +
                 " xmlns=\"urn:liberty:sa:2004-04\">" +
                 "<Status code=\"sa:OK\"/>" +
                 " <ResourceOffering entryID=\"1\">\n" +
                 " <ResourceID>http://192.168.11.2:8080/
soap/012345678901234" +
                 "</ResourceID>\n" +
                 "<ServiceInstance>\n" +
                 "<ServiceType>urn:liberty:disco:2003-08</ServiceType>\n" +
                 "<ProviderID>http://192.168.11.2:8080/soap/</ProviderID>\n"
+
                 "<Description>\n" +
                 "<CredentialRef>2sxJu9g/vvLG9sAN9bKp/8q0NKU=</
CredentialRef>\n" +
```

```
                "<SecurityMechID>urn:liberty:security:2004-04:null:Bearer</
SecurityMechID>\n" +
                " <Endpoint>http://192.168.11.2:8080/soap/IDPDS</Endpoint>\
n" +
                " </Description>" +
                "</ServiceInstance>\n" +
                " </ResourceOffering>\n" +
                "<Credentials notOnOrAfter=\"2005-11-28T18:28:44Z\">\n" +
                "<saml:Assertion xmlns:saml=\"urn:oasis:names:tc:SAML:1.0:
assertion\" AssertionID=\"2sxJu9g/vvLG9sAN9bKp/8q0NKU=\">\n" +
                "</saml:Assertion>\n" +
                "</Credentials>\n" +
                "</SASLResponse>" +
                "</s:Body>" +
                "</s:Envelope>") ;
            }
        }
        catch (Exception e)
        {}
      }
    }
    catch( Exception e)
   {}
   }

   response.setContentLength( outSoap.length() + 2 ) ;

   out.println( outSoap ) ;
}
```

Example 9-23. Dummy authentication service for `libertySample()`.

9.9 AddressBook Example

This section introduces the Java NWSF AddressBook example. The purpose of the example is to illustrate the use of Java NWSF in a way that allows developers to test and run this ID-WSF-compliant test application against a live server.

The example is a small address book application, which enables the user to locate information such as a name, address, phone number, and job title for other users. Example 9-24 below illustrates the structure of the address cards presented in XML style in order to provide an idea of what data is supported for cards. The semantics can be located in the vCard specification RFC 2426, if needed. Only a limited set of vCard fields is used in the example.

```
<Card>
    <N>
        <FAMILY/>
        <GIVEN/>
        <MIDDLE/>
    </N>
    <ADR>
        <POBOX/>
        <EXTADR/>
        <STREET/>
```

```
        <LOCALITY/>
        <REGION/>
        <PCODE/>
        <CTRY/>
    </ADR>
    <TEL type="FAX/CELL/PREF"/>
    <EMAIL/>
    <TITLE/>
    <ORG>
        <ORGNAME/>
        <ORGUNIT/>
    </ORG>
</Card>
```

Example 9-24. An example AddressBook data card.

The example utilizes a Liberty-compliant address book Web service hosted by Forum Nokia, which provides the interface specified in section 9.9.4 (AddressBook Service Description). This service is not a perfect address book service, but it illustrates well the use of NWSF. Enhancements to the service would not affect the use of the framework (indeed, they would only unnecessarily complicate this example application).

The example application is developed for Nokia Series 80 Developer Platform 2.0, Nokia 9500 and Nokia 9300 terminals. It is a J2ME application based on the following specifications:

- CDC 1.0 (see JSR-36: J2ME Connected Device Configuration)[3]
- Foundation Profile (see JSR-46: J2ME Foundation Profile)[4]
- Personal Profile (see JSR-62: J2ME Personal Profile Specification)[5]
- NWSF API 1.0

For more information on the Nokia Web Services Framework for Java API 1.0, see the Developer's Guide to Nokia Web Services Framework for Java and the Nokia Web Services Framework for Java API JavaDocs.

9.9.1 Environment Overview

Figure 9-4 below illustrates the environment of the example application and NWSF. The example application is built on top of NWSF, which uses Symbian native interfaces provided by the Service Development API. The latter framework takes care of tasks such as authentication, service discovery and connection management, allowing the application developer to focus on application-level logic.

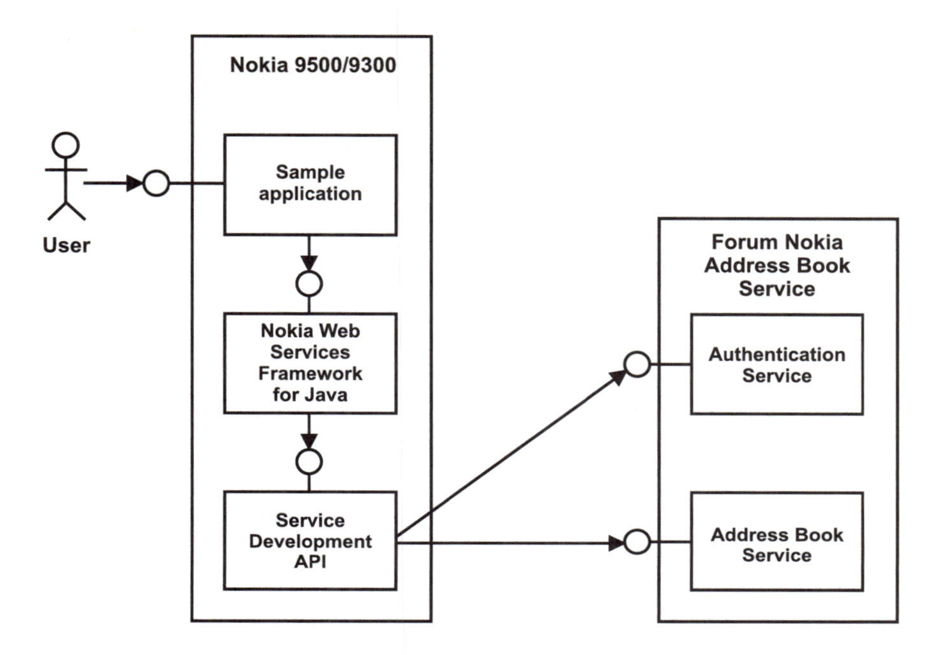

Figure 9-4. Environment of the example application.

9.9.2 Building and Running

The AddressBook example application is included in the Nokia Enhancements for Java for Series 80 enhancement package.[6] It can be downloaded from the Forum Nokia Web site (`http://forum.nokia.com/`) and installed on top of the Series 80 Developer Platform 2.0 SDK for Symbian OS – For Personal Profile. For information on how to install and configure the S80 SDK, and how to build and run the example applications, see the SDK User's Guide (included in the Series 80 Developer Platform 2.0 SDK for Symbian OS - For Personal Profile).[7]

9.9.3 Design and Implementation

This section describes the user interface (UI), application structure, user interface handling, and service handling of the example application.

9.9.3.1 User Interface

The main view of the application after it is started is illustrated in Figure 9-5. The other main UI views are the search query dialog (Figure 9-6) and search results view (Figure 9-7).

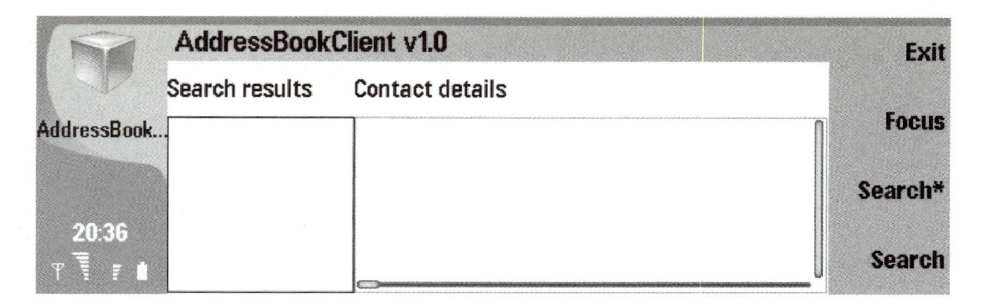

Figure 9-5. AddressBook application started.

9.9.3.2 Application Structure

The AddressBook application is a small Java application that contains four classes: `AddressBookClient`, `SearchThread`, `AddressBookFrame`, and `SearchDialog`. `AddressBookFrame` and `SearchDialog` are used for UI, `SearchThread` contains the application logic for making a search request to the address book service, and `AddressBookClient` is the main class acting as a controller between the UI and `SearchThread`.

We now present the entire application in a top-down fashion. We begin with the main controller, the `AddressBookClient`, and describe each of the other three classes in class-specific sections.

The `AddressBookClient` code appears as follows:

```
package com.nokia.mws.appl.wsf.addressbook;

/**
 * Address book client.
 */
public class AddressBookClient {
    /** AddresBookFrame for this AddressBookClient. */
    private AddressBookFrame abf = null;

    /** Thread for making the search query. */
    private SearchThread searchThread = null;

    /**
     * Main program.
     *
     * @param args Command line arguments for the main program.
     */
    public static void main(String[] args) {
        new AddressBookClient();
    }

    /** Default constructor. */
    AddressBookClient() {
        this.abf = new AddressBookFrame(this);
    }

    /**
     * Makes a search request to the address book service with given
     * queryString.
     *
```

```
     * @param queryString String to be searched.
     * @param wildcardQuery True for wildcard search, false for exact
search.
     */
    void doSearch(final String queryString, final boolean wildcardQuery) {
        if (queryString == null) {
            // No query to be made.
            return;
        }

        if (searchThread == null) {
            // Create a new SearchThread.
            searchThread = new SearchThread(this);
            searchThread.setQueryParameters(queryString, wildcardQuery);
            searchThread.start();
        }
        else {
            // Notify existing SearchThread to contine with the next search.
            searchThread.setQueryParameters(queryString, wildcardQuery);
            synchronized (searchThread) {
                searchThread.notify();
            }
        }
    }

    /**
     * Returns contact details for given contact.
     */
    String getContactDetails(String contact) {
        if (contact == null || searchThread == null) {
            return null;
        }
        return searchThread.getContactDetails(contact);
    }

    /**
     * Returns true if a search is ongoing.
     */
    boolean searchOngoing() {
        if (searchThread == null) {
            return false;
        }
        return searchThread.searchOngoing();
    }

    /**
     * Updates search results to AddressBookFrame according to search
results.
     *
     * @param statusCode Null if request was handled successfully,
     * status code string if response contained error message.
     * @param contacts Resulting contacts.
     */
    void updateSearchResultsView(String statusCode, String[] contacts) {
        abf.updateSearchResultsView(statusCode, contacts);
    }

    /**
     * Displays a message to user in AddressBookFrame contact details area.
     *
     * @param message Message to be displayed.
     */
    void displayContactDetails(String message) {
        abf.displayContactDetails(message);
    }
}
```

Example 9-25. AddressBookClient code listing.

9.9.3.3 User Interface Handling

The actual UI class is called `AddressBookFrame`, and it extends `java.awt.Frame`. The class contains the UI component definitions, and event and command button listeners. The UI layout is created in the `AddressBookFrame` constructor.

The UI is composed of the following main parts:

- Search results list for displaying the list of users matching the given search query.
- Contact details text area for displaying detailed information about the selected user. This text area is also used for displaying various status messages.
- Search dialog for the user's search string input (case-sensitive).
- Command button area for controlling the application. The main view contains the following command buttons: Exit (quits the application), Focus (changes the input focus between the search results list and contact details text area), Search* (starts a wildcard search), Search – (starts an exact search).

In addition to the common UI component definitions, the class contains two special methods. The `updateSearchResultsView` method takes the status of the search task and the search results as parameters, and in case of a successful search, updates the view with the latest search results. The `displayContactDetails` method is used to relay `String` messages to the contact details text area.

```
package com.nokia.mws.appl.wsf.addressbook;

import com.symbian.epoc.awt.CBAEvent;
import com.symbian.epoc.awt.CBAListener;
import com.symbian.epoc.awt.EikCommandButtonGroup;
import com.symbian.epoc.awt.EikStatusPane;

import java.awt.Frame;
import java.awt.GridBagConstraints;
import java.awt.GridBagLayout;
import java.awt.Label;
import java.awt.List;
import java.awt.TextArea;
import java.awt.event.ActionEvent;
import java.awt.event.ActionListener;
import java.awt.event.ItemEvent;
import java.awt.event.ItemListener;
import java.awt.event.WindowAdapter;
import java.awt.event.WindowEvent;

/**
 * Sample application user interface for accessing AddressBook Service
 * with Nokia Web Services Framework.
```

```
 */
public class AddressBookFrame extends Frame implements CBAListener {
    /** AddressBookClient for this AddressBookFrame. */
    private AddressBookClient abc = null;

    /** List of search results. */
    private List searchResultsList = null;

    /** Text area for contact details. */
    private TextArea contactDetailsTextArea = null;

    /** Command button group. */
    private EikCommandButtonGroup group = null;

    /**
     * Constructor.
     *
     * @param abclient AddressBookClient for this AddressBookFrame.
     */
    public AddressBookFrame(AddressBookClient abclient) {
        abc = abclient;

        setTitle("AddressBookClient v1.0");

        // Set status pane style.
        EikStatusPane.setStatusPaneStyle(EikStatusPane.WIDE);

        // Create UI layout.
        GridBagLayout gridbag = new GridBagLayout();
        GridBagConstraints c = new GridBagConstraints();
        setLayout(gridbag);

        c.fill = GridBagConstraints.BOTH;
        c.weightx = 1.0;
        c.weighty = 1.0;

        // Create and add UI components.
        Label searchResultsLabel = new Label("Search results");
        searchResultsList = new List();

        Label contactDetailsLabel = new Label("Contact details");
        contactDetailsTextArea = new TextArea(5, 30);
        contactDetailsTextArea.setEditable(false);

        c.gridx = 0;
        c.gridy = 0;
        gridbag.setConstraints(searchResultsLabel, c);
        c.gridx = 1;
        c.gridy = 0;
        gridbag.setConstraints(contactDetailsLabel, c);
        c.gridx = 0;
        c.gridy = 1;
        gridbag.setConstraints(searchResultsList, c);
        c.gridx = 1;
        c.gridy = 1;
        gridbag.setConstraints(contactDetailsTextArea, c);

        add(searchResultsLabel);
        add(searchResultsList);
        add(contactDetailsLabel);
        add(contactDetailsTextArea);

        // Add command buttons.
        group = new EikCommandButtonGroup();
        group.setText(EikCommandButtonGroup.BUTTON1, "Exit");
        group.setText(EikCommandButtonGroup.BUTTON2, "Focus");
```

```
        group.setText(EikCommandButtonGroup.BUTTON3, "Search*");
        group.setText(EikCommandButtonGroup.BUTTON4, "Search");
        group.addCBAListener(this);
        add(group);

        // Add event listeners.
        addWindowListener(
            new WindowAdapter() {
                public void windowClosing(WindowEvent e) {
                    System.exit(0);
                }
            });
        searchResultsList.addItemListener(
            new ItemListener() {
                public void itemStateChanged(ItemEvent e) {
                    displayContactDetails(abc.getContactDetails(
                            searchResultsList.getSelectedItem()));
                }
            });
        searchResultsList.addActionListener(
            new ActionListener() {
                public void actionPerformed(ActionEvent e) {
                    displayContactDetails(abc.getContactDetails(
                            searchResultsList.getSelectedItem()));
                }
            });

        // Display UI.
        pack();
        show();
    }

    /*
     * Handles events from command buttons.
     *
     * @see com.symbian.epoc.awt.CBAListener#cbaActionPerformed(com.symbian.
epoc.awt.CBAEvent)
     */
    public void cbaActionPerformed(CBAEvent cba) {
        if (cba.getID() == EikCommandButtonGroup.BUTTON1) {
            // Exit button.
            System.exit(0);
        }
        else if (cba.getID() == EikCommandButtonGroup.BUTTON2) {
            // Focus button
            if (contactDetailsTextArea.hasFocus()) {
                searchResultsList.requestFocus();
            }
            else {
                contactDetailsTextArea.requestFocus();
            }
        }
        else if (cba.getID() == EikCommandButtonGroup.BUTTON3) {
            // Wildcard search.
            if (!abc.searchOngoing()) {
                SearchDialog dialog =
                    new SearchDialog(this, "Wildcard search query");
                abc.doSearch(dialog.getQueryString(), true);
            }
        }
        else if (cba.getID() == EikCommandButtonGroup.BUTTON4) {
            // Exact search.
            if (!abc.searchOngoing()) {
                SearchDialog dialog =
                    new SearchDialog(this, "Exact search query");
                abc.doSearch(dialog.getQueryString(), false);
            }
```

```
        }
    }

    /**
     * Updates searchResultsList according to search results.
     *
     * @param statusCode Null if request was handled successfully,
     * status code string if response contained error message.
     * @param contacts Resulting contacts.
     */
    void updateSearchResultsView(String statusCode, String[] contacts) {
        searchResultsList.removeAll();

        if (statusCode != null) {
            displayContactDetails(statusCode);
            return;
        }

        if (contacts == null ||contacts.length == 0) {
            displayContactDetails("No results found.");
            return;
        }

        for (int i = 0; i < contacts.length; i++) {
            searchResultsList.add(contacts[i]);
        }
        displayContactDetails(abc.getContactDetails(contacts[0]));
    }

    /**
     * Displays a message to user in contactDetailsTextArea.
     *
     * @param message Message to be displayed.
     */
    void displayContactDetails(String message) {
        contactDetailsTextArea.setText(message);
    }
}
```

Example 9-26. AddressBookFrame code listing.

The name of the person being searched is entered through a dialog. The class is `SearchDialog`, and it is inherited from `java.awt.Dialog`. The class defines a query dialog brought on screen to relay user input to the actual search function, and also defines a set of command button commands used to activate or cancel the query.

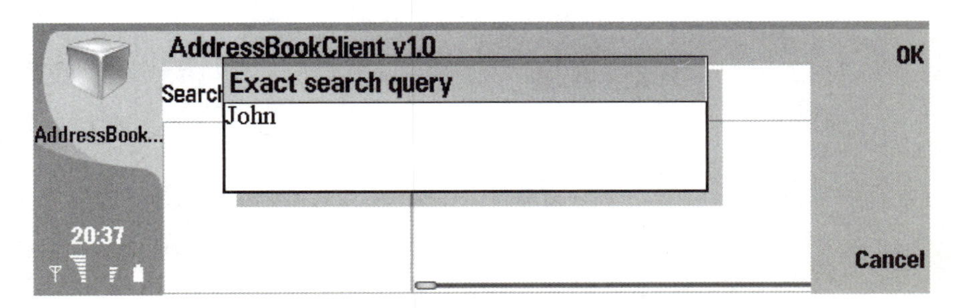

Figure 9-6. AddressBook application search dialog.

```
package com.nokia.mws.appl.wsf.addressbook;

import com.symbian.epoc.awt.CBAEvent;
import com.symbian.epoc.awt.CBAListener;
import com.symbian.epoc.awt.CKONToolkit;
import com.symbian.epoc.awt.EikCommandButtonGroup;

import java.awt.Dialog;
import java.awt.Frame;
import java.awt.Rectangle;
import java.awt.TextField;

/**
 * Dialog for requesting the query string from the user.
 */
public  class SearchDialog extends Dialog implements CBAListener {
    /** Input text field. */
    private TextField input = null;

    /** Flag telling if dialog was cancelled. */
    private boolean cancelled = false;

    /** Command button group. */
    private EikCommandButtonGroup group = null;

    /** Default constructor. */
    SearchDialog(Frame owner, String title) {
        super(owner, title, true);

        group = new EikCommandButtonGroup();
        group.setText(EikCommandButtonGroup.BUTTON1, ”OK”);
        group.setText(EikCommandButtonGroup.BUTTON4, ”Cancel”);
        group.addCBAListener(this);
        add(group);

        input = new TextField();
        add(input);

        Rectangle r = CKONToolkit.getAvailableScreenRect();
        int y = r.y;
        int x = r.x;
        int width = (int) ((3 * r.width) / 4);
        int height = r.height / 2;
        setBounds(x + (width / 8), y + (height / 4), width, height);

        setVisible(true);
        pack();
    }

    /*
     * Handles events from command buttons.
     *
     * @see com.symbian.epoc.awt.CBAListener#cbaActionPerformed(com.symbian.
epoc.awt.CBAEvent)
     */
    public void cbaActionPerformed(CBAEvent cba) {
        if (cba.getID() == EikCommandButtonGroup.BUTTON1) {
            // OK.
            dispose();
        }
        else if (cba.getID() == EikCommandButtonGroup.BUTTON2) {
            // No action.
        }
        else if (cba.getID() == EikCommandButtonGroup.BUTTON3) {
            // No action.
        }
        else if (cba.getID() == EikCommandButtonGroup.BUTTON4) {
```

```
                // Cancel.
                cancelled = true;
                dispose();
            }
        }

        /**
         * Returns the query string entered by the user, or null if the
         * dialog was cancelled.
         */
        public String getQueryString() {
            if (cancelled) {
                return null;
            }
            return input.getText();
        }
    }
```

Example 9-27. Search Dialog code listing.

9.9.3.4 Service Handling

Whereas the `AddressBookClient` class manages the whole AddressBook application, the `SearchThread` class takes care of everything relating to actual service handling. It registers services to NWSF, takes care of connections to the Web service, constructs service requests and parses service responses. Web service connections are handled and requests are processed in a separate thread – separately from the main thread. Placing all potentially blocking tasks in separate, independent threads ensures that the UI remains responsive and actions can be executed at the same time as the Web service connection and requests are being processed.

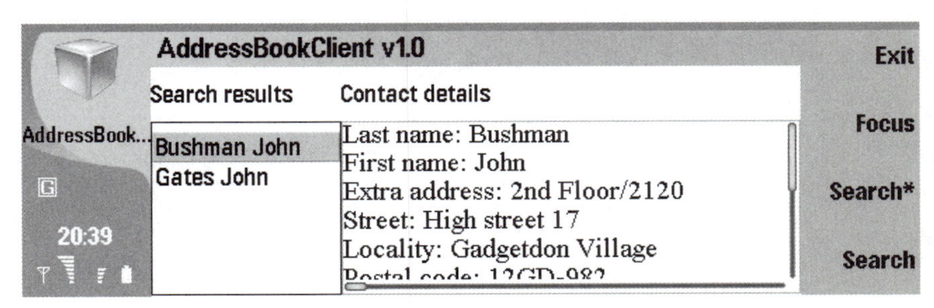

Figure 9-7. AddressBook application with search results.

The example application is closely tied to the example service hosted by Forum Nokia – all AddressBook service connection parameters are placed in the `SearchThread` class as constants. Search results are stored in a hash table called `contactDetailsMap`. The key of the hash table is a `string` containing the user's name, and the corresponding value is a `string` containing contact information to be displayed.

```
package com.nokia.mws.appl.wsf.addressbook;

import com.nokia.mws.wsf.IdentityProvider;
import com.nokia.mws.wsf.ServiceConnection;
import com.nokia.mws.wsf.ServiceDescription;
import com.nokia.mws.wsf.ServiceManager;
import com.nokia.mws.wsf.WSFException;
import com.nokia.mws.wsf.WSFObjectFactory;
import com.nokia.mws.wsf.xml.Element;
import com.nokia.mws.wsf.xml.ElementException;
import com.nokia.mws.wsf.xml.ElementFactory;

import java.util.Arrays;
import java.util.Hashtable;
import java.util.Map;

/**
 * Address book client thread for making the search request to the
 * address book service.
 */
public class SearchThread extends Thread {
    /** AddresBookClient for this SearchThread. */
    private AddressBookClient abc = null;

    /** Address book service contract. */
    private static final String ADDRESS_BOOK_CONTRACT =
        "urn:nokia:test:addrbook:2004-09";

    /** Authentication service endpoint. */
    private static final String AS_ENDPOINT =
        "http://selma.ndhub.net:9080/tfs/IDPSSO_IDWSF";

    /** Authentication service contract. */
    private static final String AS_CONTRACT = "urn:liberty:as:2004-04";

    /** Authentication service provider id. */
    private static final String AS_PROVIDER_ID =
        "http://selma.ndhub.net:9080/tfs/";

    /** Advisory authentication id. */
    private static final String TEST_ADVISORY_AUTHN_ID = "testuser1";

    /** Authentication password. */
    private static final String TEST_PASSWORD = "testuser1";

    /** True when search request/response processing is ongoing. */
    private boolean searchOngoing = false;

    /** Map containing contact details. */
    private Map contactDetailsMap = null;

    /** ServiceConnection to the address book service. */
    private ServiceConnection serviceConnection = null;

    /** String to be searched. */
    private String queryString = null;

    /**
     * Constructor.
     *
     * @param abclient AddressBookClient for this AddressBookFrame.
     */
    SearchThread(AddressBookClient abclient) {
```

```
        abc = abclient;
        contactDetailsMap = new Hashtable();
    }
    /** True for wildcard search, false for exact search. */
    private boolean wildcardQuery = true;
```

Example 9-28. SearchThread code listing.

The Web service connection is handled by two methods: `initializeServices()` and `getServiceConnection()`. The `initializeServices()` method instantiates IDP and service description objects with necessary data such as endpoint and login information. The exact composition of the required information is visible in Table 9-1 below. This method only needs to be executed once, since NWSF for Java stores IDP and service description information persistently.

PARAMETER NAME	PARAMETER VALUE
Service Contract	urn:nokia:test:addrbook:2004-09
Authentication Service Endpoint	http://selma.ndhub.net:9080/tfs/IDPSSO_IDWSF
Authentication Service Contract	urn:liberty:as:2004-04
Authentication Service Provider ID	http://selma.ndhub.net:9080/tfs/
Advisory Authentication ID	testuser1
Password	testuser1
WsfFrameworkId	ID-WSF

Table 9-1. AddressBook service connection parameters.

```
    /**
     * Registers an identity provider and the address book service and
     * associates them with each other. This needs to be done only
     * once.
     */
    private void initializeServices() throws WSFException {
        ServiceManager serviceManager = WSFObjectFactory.getInstance()

 .getServiceManager();

        IdentityProvider idp =
            WSFObjectFactory.getInstance().getIdentityProvider(
                AS_PROVIDER_ID, AS_ENDPOINT, AS_CONTRACT, null);
        idp.setUserInfo(null, TEST_ADVISORY_AUTHN_ID, TEST_PASSWORD);

        ServiceDescription sd =
            WSFObjectFactory.getInstance().getServiceDescription(
                null, ADDRESS_BOOK_CONTRACT, null);

        serviceManager.registerIdentityProvider(idp);
        serviceManager.registerServiceDescription(sd);
        serviceManager.associateService(sd, idp);
    }
```

Example 9-29. Creating an IDP–Service Provider association.

The `getServiceConnection()` method instantiates a service description object using an address book service contract, and obtains a service connection instance from the Service Manager. It returns the service connection that the application uses to submit Web service requests to the address book service.

```
    /**
     * Returns service connection for the address book service.
     */
    private ServiceConnection getServiceConnection() throws WSFException {
        if (serviceConnection == null) {
            initializeServices();

            ServiceDescription sd =
                WSFObjectFactory.getInstance().getServiceDescription(
                    null, ADDRESS_BOOK_CONTRACT, null);

            serviceConnection =
                WSFObjectFactory.getInstance().getServiceManager()
                    .getServiceConnection(sd, null);
        }

        return serviceConnection;
    }
```

Example 9-30. Establishing a service connection.

The `run` method of the `SearchThread` class makes the actual search request. It obtains a service connection, constructs a request, submits the request, and gets a response. Finally, it parses the response and displays the results. After each completed request, the thread waits until it is activated to process another search request.

```
/**
     * Makes a search request to the address book service with given
     * query parameters. Application uses one thread for making all the
     * requests, AddressBookClient notifies thread when the next search
     * needs be started.
     */
    public void run() {
        while (true) {
            try {
                searchOngoing = true;
                if (serviceConnection == null) {
                    abc.displayContactDetails("Connecting...");
                    serviceConnection = getServiceConnection();
                }
                String queryRequest =
                    getQueryRequest(queryString, wildcardQuery);
                abc.displayContactDetails("Submitting request...\n"
                    + (wildcardQuery ? "Wildcard search: ": "Exact search:
")
                    + queryString);
                String queryResponse =
                    serviceConnection.submit(queryRequest);
                abc.displayContactDetails("Parsing response...");
                String statusCode = parseQueryResponse(queryResponse);
                String[] contacts =
                    (String[])contactDetailsMap.keySet()
                        .toArray(new String[0]);
```

```
                Arrays.sort(contacts);
                abc.updateSearchResultsView(statusCode, contacts);
            }
            catch (WSFException wsfe) {
                abc.displayContactDetails(wsfe.toString());
                // Something went wrong in request sending, obtain a new
                // connection for the next request.
                serviceConnection = null;
            }
            catch (ElementException ee) {
                abc.displayContactDetails(ee.toString());
            }
            finally {
                try {
                    searchOngoing = false;
                    synchronized (this) {
                        wait();
                    }
                }
                catch (InterruptedException ie) {
                    abc.displayContactDetails(ie.toString());
                }
            }
        }
    }

    /**
     * Returns true if a search is ongoing.
     */
    boolean searchOngoing() {
        return searchOngoing;
    }

    /**
     * Set parameters for the next query to be made.
     *
     * @param queryString String to be searched.
     * @param wildcardQuery True for wildcard search, false for exact
search.
     */
    void setQueryParameters(String queryString, boolean wildcardQuery) {
        this.queryString = queryString;
        this.wildcardQuery = wildcardQuery;
    }

    /**
     * Returns contact details for given contact.
     */
    String getContactDetails(String contact) {
        if (contact == null) {
            return null;
        }
        return (String)contactDetailsMap.get(contact);
    }
```

Example 9-31. Creating a search request.

Web service requests and responses are handled by two methods: `getQueryRequest()` and `parseQueryResponse()`, respectively. The `getQueryRequest()` method constructs the search query request to be sent to the address book service. The following example describes a basic query to request all address cards containing the phone number 1234.

```
<ab:Query xmlns:ab="urn:nokia:test:addrbook:2004-09">
<ab:ResourceID>XYZ</ab:ResourceID>
<ab:QueryItem>
        <ab:Select>/ab:Card[ab:TEL="1234"]</ab:Select>
</ab:QueryItem>
</ab:Query>
```

Example 9-32. Sample query searching for cards with the phone number "1234".

When comparing the example query with any arbitrary example, the only parts changing are the `<ResourceID>` and `<Select>` parameters. The `<Select>` parameter is a string defining what is requested. It contains an XPath expression based on XPath 1.0. Expressions that can be used on the client side include, for example, the following:

- Querying based on the phone number: `/ab:Card[ab:TEL="1234"]`
- Querying based on the family name: `/ab:Card[ab:N/ab:FAMILY="Smith"]`
- Querying based on the given name: `/ab:Card[ab:N/ab:GIVEN="John"]`
- Querying based on the family and given name: `/ab:Card[ab:N/ab:FAMILY="Smith" and ab:N/ab:GIVEN="John"]`

In the example application, the request is constructed using the `com.nokia.mws.wsf.xml.Element` interface, which is also used for parsing the response. However, it could just as well be constructed as a string, or by any other means of XML generation. The request selects those users whose family name, given name, or telephone number contains the query string from the address book.

```
    /**
     * Returns an XML request to be sent to the address book service
     * when given queryString is to be searched.
     *
     * @param queryString String to be searched.
     * @param wildcardQuery True for wildcard search, false for exact
search.
     * @return XML request to be sent to the address book service.
     */
    private String getQueryRequest(String queryString, boolean
wildcardQuery)
            throws ElementException {

        // Construct selection string (XPath 1.0 expression).
        String selectionString = null;
        if (wildcardQuery) {
            selectionString =
                "/ab:Card[contains(ab:N/ab:FAMILY,\"" + queryString
                + "\") or contains(ab:N/ab:GIVEN, \"" + queryString
                + "\") or contains(ab:N/ab:TEL, \"" + queryString
                + "\")]";
```

```
        }
        else {
            selectionString =
                "/ab:Card[ab:TEL=\"" + queryString
                + "\" or ab:N/ab:FAMILY=\"" + queryString
                + "\" or ab:N/ab:GIVEN=\"" + queryString
                + "\"]";
        }

        // Use Element interface to construct the XML document.
        Element request = ElementFactory.getInstance().createElement(
            "ab", "Query", "urn:nokia:test:addrbook:2004-09");
        request.addElement("ResourceID").setContent("XYZ");
        request.addElement("QueryItem").addElement("Select")
            .setContent(selectionString);
        return request.toXml();
    }
```

Example 9-33. Creating an XML query.

The response has two changing parts: a status code attribute and the actual returned data. The status code depends on the processing results on the service side, either informing the client that everything went smoothly, or that something went wrong. The `<Data>` element contains the returned data. It can include one or more `<Card>` elements housing the actual contact details.

```
<ab:QueryResponse xmlns:ab="urn:nokia:test:addrbook:2004-09">
<ab:Status code="OK"/>
<ab:Data>
<ab:Card>
<ab:N>
<ab:FAMILY>Smith</ab:FAMILY>
<ab:GIVEN>John</ab:GIVEN>
</ab:N>
<ab:ADR>
<ab:EXTADR>Room 123</ab:EXTADDR>
<ab:STREET>1st ST 123</ab:STREET>
<ab:LOCALITY>Huitsi</ab:LOCALITY>
<ab:REGION>Nevada</ab:REGION>
<ab:PCODE>7698532</ab:PCODE>
</ab:ADR>
<ab:TEL>1234</ab:TEL>
<ab:TEL type="CELL">2122345</ab:TEL>
<ab:EMAIL>john.smith@acme.com</ab:EMAIL>
<ab:TITLE>Gadget Manager</ab:TITLE>
</ab:Card>
</ab:Data>
</ab:QueryResponse>
```

Example 9-34. A query response in XML.

The `parseQueryResponse()` method takes a response string returned by the address book service and parses it using the `com.nokia.mws.wsf.xml.Element` interface. It adds the users and their details into the `contactDetailsMap` hash table, from which they are fetched to be displayed.

```
/**
     * Parses query response from the address book service and updates
     * contactDetailsMap with new contacts and their details.
     *
     * @param queryResponse Response from the address book service.
     * @return Null if request was handled successfully, status code
     * string if response contained error message.
     */
    private String parseQueryResponse(String queryResponse)
            throws ElementException {
        contactDetailsMap.clear();

        String contact = "";
        String contactDetails = "";
        String statusCode = null;
        Element e = ElementFactory.getInstance().parseElement(queryResponse
);

        while (e != null) {
            if (e.getName().equals("Card")) {
                // New card begins.
                if (contact.length() > 0) {
                    // Add previous contact to contactDetailsMap.
                    contactDetailsMap.put(contact, contactDetails);
                    contact = "";
                    contactDetails = "";
                }
            }
            else if (e.getName().equals("Status")) {
                // Check status of the operations.
                String code = e.getAttributeValue("code");
                if (!code.equalsIgnoreCase("OK")) {
                    if (statusCode == null) {
                        statusCode = "Status: " + code;
                    }
                    else {
                        statusCode += ", " + code;
                    }
                }
            }
            else if (e.getName().equals("FAMILY")) {
                contact = e.getContent();
                contactDetails = "Last name: " + e.getContent() + "\n";
            }
            else if (e.getName().equals("GIVEN")) {
                contact += (" " + e.getContent());
                contactDetails += ("First name: " + e.getContent() + "\n");
            }
            else if (e.getName().equals("MIDDLE")) {
                contact += (" " + e.getContent());
                contactDetails += ("Middle name: " + e.getContent() + "\n");
            }
            else if (e.getName().equals("POBOX")) {
                contactDetails += ("PO box: " + e.getContent() + "\n");
            }
            else if (e.getName().equals("EXTADR")) {
                contactDetails += ("Extra address: " + e.getContent() +
"\n");
            }
            else if (e.getName().equals("STREET")) {
                contactDetails += ("Street: " + e.getContent() + "\n");
            }
            else if (e.getName().equals("LOCALITY")) {
                contactDetails += ("Locality: " + e.getContent() + "\n");
            }
            else if (e.getName().equals("REGION")) {
```

```
                contactDetails += ("Region: " + e.getContent() + "\n");
            }
            else if (e.getName().equals("PCODE")) {
                contactDetails += ("Postal code: " + e.getContent() + "\n");
            }
            else if (e.getName().equals("CTRY")) {
                contactDetails += ("Country: " + e.getContent() + "\n");
            }
            else if (e.getName().equals("EMAIL")) {
                contactDetails += ("Email: " + e.getContent() + "\n");
            }
            else if (e.getName().equals("TITLE")) {
                contactDetails += ("Title: " + e.getContent() + "\n");
            }
            else if (e.getName().equals("ORGNAME")) {
                contactDetails += ("Organization: " + e.getContent() + "\
n");
            }
            else if (e.getName().equals("ORGUNIT")) {
                contactDetails += ("Unit: " + e.getContent() + "\n");
            }
            else if (e.getName().equals("TEL")) {
                if ("PREF" == e.getAttributeValue("type")) {
                    contactDetails += ("Phone: " + e.getContent() + "\n");
                }
                else if ("CELL" == e.getAttributeValue("type")) {
                    contactDetails += ("Mobile phone: " + e.getContent() +
"\n");
                }
                else if ("FAX" == e.getAttributeValue("type")) {
                    contactDetails += ("Fax: " + e.getContent() + "\n");
                }
            }

            //Get the next element.
            e = e.getNextElement();
        }

        if (contact.length() > 0) {
            // Add the last contact to contactDetailsMap.
            contactDetailsMap.put(contact, contactDetails);
        }

        return statusCode;
    }
```

Example 9-35. Parsing a query response.

9.9.4 AddressBook.wsdl

```
<definitions
 xmlns="http://schemas.xmlsoap.org/wsdl/"
 xmlns:soap="http://schemas.xmlsoap.org/wsdl/soap/"
 xmlns:xs="http://www.w3.org/2001/XMLSchema"
 xmlns:soapenc="http://schemas.xmlsoap.org/soap/encoding/"
 xmlns:ab="urn:nokia:test:addrbook:2004-09"
 targetNamespace="urn:nokia:test:addrbook:2004-09">
  <import namespace="urn:liberty:as:2004-04" location="liberty-idwsf-authn-
svc-v1.0.wsdl"/>
  <types>
    <xs:schema>
      <xs:import schemaLocation="AddrBook.xsd"/>
    </xs:schema>
  </types>
```

```
  <message name="Query">
    <part name="body" element="ab:Query"/>
  </message>
  <message name="QueryResponse">
    <part name="body" element="ab:QueryResponse"/>
  </message>
  <portType name="AddressBookPort">
    <operation name="AddressBookQuery">
      <input message="ab:Query"/>
      <output message="ab:QueryResponse"/>
    </operation>
  </portType>
  <binding name="AddressBookBinding" type="ab:AddressBookPort">
    <!-- ext:WSDLBinding>urn:liberty:sb:2003-08</ext:WSDLBinding -->
    <soap:binding style="document" transport="http://schemas.xmlsoap.org/
soap/http"/>
    <operation name="AddressBookQuery">
      <input>
        <soap:body use="literal"/>
      </input>
      <output>
        <soap:body use="literal"/>
      </output>
    </operation>
  </binding>
  <service name="AddressBookService"/>
</definitions>
```

9.9.5 AddrBook.xsd

```
<?xml version="1.0" encoding="UTF-8"?>
<xs:schema
 targetNamespace="urn:nokia:test:addrbook:2004-09"
 xmlns="urn:nokia:test:addrbook:2004-09"
 xmlns:xs="http://www.w3.org/2001/XMLSchema"
 elementFormDefault="qualified">
  <xs:include schemaLocation="liberty-idwsf-dst-v1.0.xsd"/>
  <xs:simpleType name="SelectType">
    <xs:restriction base="xs:string"/>
  </xs:simpleType>
  <xs:element name="N">
    <xs:complexType>
      <xs:sequence>
        <xs:element name="FAMILY" type="xs:string" minOccurs="0"/>
        <xs:element name="GIVEN" type="xs:string" minOccurs="0"/>
        <xs:element name="MIDDLE" type="xs:string" minOccurs="0"/>
      </xs:sequence>
    </xs:complexType>
  </xs:element>
  <xs:element name="ADR">
    <xs:complexType>
      <xs:sequence>
        <xs:element name="POBOX" type="xs:string" minOccurs="0"/>
        <xs:element name="EXTADR" type="xs:string" minOccurs="0"/>
        <xs:element name="STREET" type="xs:string" minOccurs="0"/>
        <xs:element name="LOCALITY" type="xs:string" minOccurs="0"/>
        <xs:element name="REGION" type="xs:string" minOccurs="0"/>
        <xs:element name="PCODE" type="xs:string" minOccurs="0"/>
        <xs:element name="CTRY" type="xs:string" minOccurs="0"/>
      </xs:sequence>
    </xs:complexType>
  </xs:element>
  <xs:element name="TEL">
    <xs:complexType>
```

```
      <xs:simpleContent>
        <xs:extension base="xs:string">
          <xs:attribute name="TYPE">
            <xs:simpleType>
              <xs:restriction base="xs:string">
                <xs:enumeration value="FAX"/>
                <xs:enumeration value="CELL"/>
                <xs:enumeration value="PREF"/>
              </xs:restriction>
            </xs:simpleType>
          </xs:attribute>
        </xs:extension>
      </xs:simpleContent>
    </xs:complexType>
  </xs:element>
  <xs:element name="EMAIL" type="xs:string"/>
  <xs:element name="TITLE" type="xs:string"/>
  <xs:element name="ORG">
    <xs:complexType>
      <xs:sequence>
        <xs:element name="ORGNAME" type="xs:string"/>
        <xs:element name="ORGUNIT" type="xs:string" minOccurs="0"
maxOccurs="4"/>
      </xs:sequence>
    </xs:complexType>
  </xs:element>
  <xs:element name="Card">
    <xs:complexType>
      <xs:sequence>
        <xs:element ref="N"/>
        <xs:element ref="ADR" minOccurs="0"/>
        <xs:element ref="TEL" minOccurs="0" maxOccurs="4"/>
        <xs:element ref="EMAIL" minOccurs="0"/>
        <xs:element ref="TITLE" minOccurs="0"/>
        <xs:element ref="ORG" minOccurs="0"/>
      </xs:sequence>
    </xs:complexType>
  </xs:element>
</xs:schema>
```

Chapter 10: C++ Client Development

This chapter introduces the Nokia Service Development API, and describes the underlying framework (called Nokia Web Services Framework for Series 80 (NWSF)). While the architecture and key concepts of NWSF are described in chapter 7, this chapter focuses on code examples illustrating the concepts described in chapter 7.

In order to provide practical examples illustrating the actual usage of the APIs, we present three sample applications. The main example used in this document is a "Hello World" Web service consumer (WSC) application (HelloWS) that consists of a set of C++ classes. Some classes are generated automatically (skeleton code) by transforming XML-based Web service description and schema definitions into C++ code and project definition files. This demonstrates how developers can efficiently utilize the NWSF APIs. The example provides a better understanding of the classes that are important to application-specific processing (as opposed to the communications framework).

Another, more advanced sample code, is the "PhonebookEx" application distributed with the Series 80 C++ SDK and located in the directory `\epoc32\Series80Ex\`. This simple address book example presents an ID-WSF-enabled WSC that can be tested against a live service provider. We want to acknowledge and thank Trustgenix, Inc., which implemented a test service provider on the `http://selma.ndhub.net:9080/tfs` *Internet access point (IAP).*

The third sample application called wscexample is the most advanced one, and functions as the main example of this chapter. The application contains several features, and in effect combines the features of both of the C++ examples described above into a single package. The application utilizes a set of test servlets (covered in section 10.12) in order to provide basic messaging features and secure messaging via HTTP Basic Authentication. The application also provides console-based access the to ID-WSF-based AddressBook service covered in the PhonebookEx example. Unlike the Java example, this application does not contain the dummy ID-WSF example, because we are able to use the live AddressBook service to demonstrate the same features. Relevant snippets of the source code are presented with comments in section 10.11. Although this book concentrates on Series 80 software development, the sample application contains several comments that also apply to the SOA for S60 platform APIs. With the information presented in the comments, it is easy to port the application to the SOA for S60 platform.

NOTE
Please check the Forum Nokia Web site and the present book's Web page on the publisher's Web site (http://www.wiley.com) for source-code downloads.

For demonstration purposes, the PhonebookEx application is coded directly against the NWSF APIs. Another approach to developing Web service applications is to use tools that generate skeleton code based on XML schema and WSDL descriptions. These tools generalize the approach demonstrated in the HelloWS example, as described above. Nokia provides one such tool that generates C++ code stubs as an add-in to Microsoft Visual Studio. The tool is described in detail in the Nokia WSDL-to-C++ Wizard User's Guide.[WSDL2C]

While this chapter discusses the APIs used to create WSC applications, it does not cover the creation of Web service provider (WSP) applications. Nonetheless, the collection of Java servlets for the HelloWS service described in Section 9.8.1 emulates actual Web service servers. This also applies to the "Hello World" Web service consumer example written in Symbian C++.

This chapter is aimed at Symbian developers who are already familiar with the Symbian development environment, particularly with Symbian Active Objects. In addition, some familiarity with the general concepts related to Web Services and the Liberty Alliance specifications is recommended.[LIBERTY]

Note that the features described in this chapter require the installation of the Nokia platform SDKs and the necessary update packages. Please refer to the installation instructions for details.

CASE STUDY: RADIO@AOL

The Radio@AOL is the result of a joint proof-of-concept project between AOL, Nokia, and Trustgenix. With a project time of approximately six weeks, the team was able to create a mobile version of the solution using Liberty ID-WSF specifications. The client is written in C++, and it utilizes the SOA for S60 platform. One important aspect of the demo was to prove how generic protocols could be used also in mobile use cases.

The business model could be based on an agreement between the service provider and an operator, where the operator would be responsible for providing and charging for wireless data access to users.

Figure 10-1 below illustrates the case on a high level. It is assumed that the client has already been provisioned to the user, either as a pre-installed component or through user registration and OTA download.

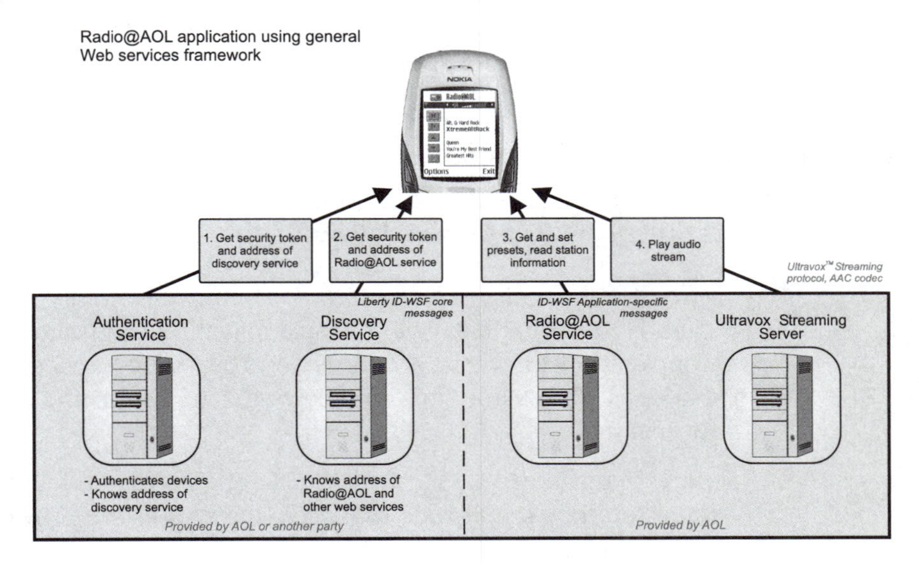

Figure 10-1. Radio@AOL application using general Web services framework.

The user activates the client with one click. Immediately after the activation, the client automatically performs authentication; in this case, using the *International Mobile Equipment Identity (IMEI)* number. Once confirmed, the Authentication Service (AS) returns a security token and the address to the Discovery Service (DS).

The client then – again automatically – contacts the Discovery Service and in the response message receives a new security token and the address of Radio@AOL.

This step might seem unwarranted, but one should bear in mind the potential scalability of the services: the client might support several services, and the role of the DS could be to provide a list of all available services.

It is the time to invoke the Radio@AOL service. If the user has pre-set a favorite channel in the past, the request is to read (again automatically) the station information of the preferred channel. (The user can, of course, interrupt this process at any time and select another channel.)

Finally, the Radio@AOL audio stream is started and the user can enjoy the music.

10.1 Nokia Service Development API

The Nokia Service Development API is a set of dynamic link libraries that allow application developers to write Web service applications. Currently support is provided for WSC applications, but in the future, it will also be possible to write WSP applications that run on Nokia mobile devices. The Service Development API provides a comprehensive set of APIs that allow developers to write service consumer applications based on SOAP, Web Services Security, Web Services Interoperability, and the Liberty Alliance Identity Web Services Framework (ID-WSF). [SOAP], [WSS], [WS-I] In addition, some functions are provided to support XML processing and other utility applications.

10.1.1 General Package Structure

The Service Development API consists of a number of dynamic link libraries. These are structured as follows:

1. Service Connection (`SENSERVCONN.lib`). This library contains a class for creating a NWSF client application, and allows such an application to create a connection to a Web service. It also includes a callback interface for receiving status, error, and message information.
2. Service Manager (`SENSERVMGR.lib`). The NWSF architecture is administered by the Service Manager, and classes are provided to enable client applications to interact with the Service Manager (for example, to register services and identity provider information).
3. Service Description (`SENSERVDESC.lib`). This set of classes for managing service descriptions, including the creation and manipulation of service descriptions.
4. XML Processing (`SENXMLINTERFACE.lib` and `SENXML.lib`). This is a set of basic classes for XML processing. The classes include `SenBaseFragment` and `SenBaseElement`, which can be used to ease XML processing (they provide simple API for XML (SAX) and DOM-like interfaces to XML).
5. Utilities (`SENUTILS.lib`). This set of classes provides utility functions such as SOAP processing, date utilities, and GUID generation. The five libraries included in this set are further described in the subsequent sections.

10.1.2 IDE Usage

The Service Development API can be used with any Symbian development environment (for example, MS Visual Studio, Borland BuilderX, Eclipse, or Emacs). Header files and libraries are included as usual in the Symbian project files. Please note that the separate WSDL-based code generation tool plug-in is available only for the MS Visual Studio environment.

10.1.3 NWSF XML Library

NWSF is highly dependent on XML processing. In general, NWSF client applications explicitly use XML parsing when either receiving or sending content as XML. In this book, C++-based XML processing is divided into two parts. Section 7.7 provides a general introduction to XML processing, and describes the SOA for S60 platform APIs in detail. This section presents the respective NWSF for Series 80 APIs. The detailed UML schematics of the classes (among other things) are presented in the Appendix B.

The underlying platform provides basic SAX parsing capabilities. In addition, NWSF offers the `SenBaseFragment` and `SenBaseElement` classes to simplify the operations noted above. Some examples in this chapter use either these classes directly, or classes (such as `SenSoapMessage`) that are derived from these classes. The basic difference between `SenBaseFragment` and `SenBaseElement` is that the former relies on SAX parsing (i.e., the provision of event handlers for the various SAX events) while the latter provides a simple DOM interface.

```
CSenElement* helloString = CSenBaseElement::NewL(_L("HelloString"));

CSenElement* helloRequest = '
CSenBaseElement::NewL(_L("urn:nokia:ws:samples"), _L("RequestHello"));

CSenSoapMessage* soapRequest = CSenSoapMessage::NewL() ;

CleanupStack::PushL(helloString);
CleanupStack::PushL(helloRequest);
CleanupStack::PushL(soapRequest) ;

helloString->SetContentL( requestString ) ;
helloRequest->SetContentL( *(helloString->AsXmlL()) ) ;
```

Example 10-1. Using XMLElement to create a new XML-based type.

NOTE

An XML-based type can be created in roughly the same way using the `SenBaseFragment` class too. It is possible to get fragment content as an element using the `AsElement` and `ExtractElement` (the method caller takes ownership of the element) methods available in the `SenBaseFragment` class.

It is possible to use a custom content handler (`HandleMessage`) to handle incoming messages (see section 10.3.2), even though this can also be re-implemented by using `SenBaseFragment`. The latter alternative is simpler if the content of the incoming message is known (for example, if you have made a synchronous request to a service and are expecting a particular message in response):

```
CSenBaseFragment* message = CSenBaseFragment::NewL("HelloResponse") ;
message->BuildFrom( aMessage ) ;
```

Example 10-2. Using BaseFragment to parse an incoming message.

It is then possible to select individual pieces of information from the fragment by first getting content as element data (by using `ExtractElement()`) and then asking for the contents of a particular attribute or element (by using `Element( localName )` or `AttrValue( name )`).

10.1.4 WSDL and NWSF

By using XML schema files and WSDL documents to describe Web service interfaces, it is possible to quickly generate client code that can be used to access a Web service. Several integrated development environment (IDE) products allow you to generate consumer code from a WSDL file. This can accelerate the development of service consumer applications.

Nokia provides a WSDL-based code generation tool for the MS Visual Studio environment. The tool enables the creation of a consumer application stub that uses methods offered in the Service Development API. For more information on the tool plug-in, please visit the Forum Nokia Web site.

When transforming a WSDL document into a Symbian NWSF service consumer application, three transformations should be taken into account:

1. Messages described in the service ports are transformed into `SenBaseFragment` classes.
2. Types that are referred to by these messages can also be transformed into `SenBaseFragment` classes.

3. If the service is based on a third-party identity provider model (such as Liberty ID-WSF), then a WSDL document may contain a reference to the WSDL documents pertaining to an identity provider (IDP) or a discovery service provider. These WSDL documents can be used to provision NWSF with the service providers, and to associate them with the services described in the main WSDL document.

4. NWSF allows the use of multiple frameworks (and of no framework at all) to access Web services. Therefore, it is possible to select a framework based on the WSDL document, or the selection of an option. A WSDL document is an XML file that should be transformed into Symbian C++ code:

```
<?xml version="1.0"?>
<definitions xmlns="http://schemas.xmlsoap.org/wsdl/"
 xmlns:xs="http://www.w3.org/2001/XMLSchema"
 xmlns:soap="http://schemas.xmlsoap.org/wsdl/soap/"
 xmlns:ext="urn:liberty:sb:2003-08"
 targetNamespace="urn:nokia:samples:helloWS:2005-01">
  <import namespace="urn:liberty:as:2004-04"
   location="liberty-idwsf-authn-svc-v1.0.wsdl"/>
  <types>
    <xs:schema>
      <xs:import namespace="urn:nokia:samples:helloWS:2005-01"
       schemaLocation="HelloWS.xsd"/>
    </xs:schema>
  </types>
  <message name="Hello">
    <part name="body" element="RequestHello"/>
  </message>
  <message name="HelloResponse">
    <part name="body" element="HelloResponse"/>
  </message>
  <portType name="HelloWSPort">
    <operation name="HelloWSQuery">
      <input message="Hello"/>
      <output message="HelloResponse"/>
    </operation>
  </portType>
  <binding name="HelloWSBinding" type="HelloWSPort">
    <soap:binding style="document"
     transport="http://schemas.xmlsoap.org/soap/http"/>
    <operation name="HelloWSQuery">
      <input>
        <soap:body use="literal"/>
      </input>
      <output>
        <soap:body use="literal"/>
      </output>
    </operation>
  </binding>
 <service name="HelloWS">
    <port name="HelloWSPort">
      <soap:address location="http://localhost:8080/soap/HelloWS"/>
    </port>
 </service>
</definitions>
```

Example 10-3. HelloWS.wsdl document that can be used to create Symbian C++ code.

Although this simple example does not map XML types into `SenBaseFragment` classes, it does generate the actual messages and client code needed to create a service consumer application. The WSDL document contains very little XSLT code: it is mostly boilerplate code. The following pieces of the WSDL document are particularly important:

```
<xs:import namespace="urn:nokia:hello-dst:2003-08"
 schemaLocation="nokia-liberty-idwsf-dst-v1.0.xsd"/>
<xs:import namespace="urn:nokia:samples:helloWS:2005-01"
 schemaLocation="HelloWS.xsd"/>
```

Example 10-4. Referencing two specific services in a WSDL document.

The lines described in Example 10-4 reference two other services, which can be registered (and possibly associated) with the service described in the WSDL document. In a WSDL file, it is often difficult to distinguish between a service that relies on the Liberty ID-WSF framework and a service that does not rely on any particular framework, but instead defines SOAP headers that can be sent by a requester. When generating NWSF code, this distinction should be made because a framework-based application does not have to submit full SOAP envelopes or add SOAP headers to a message before sending it. To make it easy for the WSDL-to-C++ tool to make this distinction, you need to add either an extension element to the WSDL document (allowed by the WSDL schema) or a command-line parameter to the XSLT transformation command itself. In the example, an element is added to the `<service>` definition:

```
<binding name="HelloWSBinding" type="HelloWSPort">
  <ext:WSDLBinding>urn:liberty:sb:2003-08</ext:WSDLBinding>
  ...
</binding>
```

Example 10-5. A WSDL extension element.

10.2 Hello Web Service Example

The following example introduces the NWSF APIs. There are generally two entities involved in a Web service interaction: a WSP and a WSC. The example starts by introducing the Web service provider – in this case, a service provider that responds to a request with a message stating "Hello Web Service Consumer!" The messages sent between the WSC and the WSP can be defined by the following XML schema document:

```
<?xml version="1.0" encoding="UTF-8" ?>
<xs:schema targetNamespace="urn:nokia:samples:helloWS:2005-01"
xmlns:xs="http://www.w3.org/2001/XMLSchema"
    xmlns="urn:nokia:samples:helloWS:2005-01" version="1.1-01">
```

```
    <xs:complexType name="RequestHelloType">
      <xs:sequence>
        <xs:element name="HelloString" type="xs:string" />
      </xs:sequence>
    </xs:complexType>
    <xs:element name="RequestHello" type="RequestHelloType" />
    <xs:complexType name="HelloResponseType">
      <xs:sequence>
        <xs:element name="ResponseString" type="xs:string" />
      </xs:sequence>
    </xs:complexType>
    <xs:element name="HelloResponse" type="HelloResponseType" />
</xs:schema>
```

Example 10-6. An XML schema document describing HelloWS messages.

Together with WSDL, the XML schema document provides the necessary input data for code generation tools.

In this simple example, the WSC sends a `<RequestHello>` message to the Web service. The corresponding SOAP request document (sent over HTTP) is illustrated in Example 10-7 below.

```
POST /soap/HelloWS HTTP/1.1
Host:
Accept: text/xml
Expect: 100-continue
User-Agent: NWSF
Content-Length: 203
Content-Type: text/xml
SOAPAction:
<S:Envelope xmlns:S="http://schemas.xmlsoap.org/soap/envelope/">
  <S:Body>
    <RequestHello xmlns="urn:nokia:ws:samples">
      <HelloString>Hello Web Service Provider!</HelloString>
    </RequestHello>
  </S:Body>
</S:Envelope>
```

Example 10-7. NWSF WSC sends a Hello Request.

The WSP responds with a `<HelloResponse>` message:

```
HTTP/1.1 100 Continue
HTTP/1.1 200 OK
Date: Thu, 23 Sep 2004 19:57:39 GMT
Server: Jetty/4.2.21 (Windows 2000/5.0 x86 java/1.4.2_04)
Content-Type: text/xml
Content-Length: 270
<S:Envelope xmlns:S="http://schemas.xmlsoap.org/soap/envelope/">
  <S:Body>
    <ex:HelloResponse xmlns:ex="urn:nokia:ws:samples">
      <ex:ResponseString>
        Hello Web Service Consumer!
      </ex:ResponseString>
    </ex:HelloResponse>
  </S:Body>
</S:Envelope>
```

Example 10-8. HelloWS Web service responds.

The message flow in these SOAP documents was generated with the following client code:

```
void CHelloWSAppUi::HandleCommandL(TInt aCommand)
{
    _LIT(title,"Web Service Information");
     switch (aCommand)
     {
     case EEikCmdExit:
         {
             CBaActiveScheduler::Exit();
             break;
         }
     case EEditServiceCmd:
         {
             // TODO
             break;
         }
     case EConnectCmd:
         {
             iService->ConnectToServiceL();
             TBool aReady;
             iService->GetConnection()->IsReady(aReady);
             if (aReady)
             {
                _LIT(msgOK,"Connect Request: Successful");
                CCknInfoDialog::RunDlgLD(title, msgOK);
             } else {
                _LIT(msgNone,"Not Connected - Status:");
                TBuf<64> buf(msgNone);
                TInt status = iService->GetStatus();
                buf.Num(status);
                CCknInfoDialog::RunDlgLD(msgNone, buf);
             }
             break;
         }
     case EEditMessageCmd:
         {
             // TODO
             break;
         }
     case ESendCmd:
         {
             _LIT(KWSRequestString, "Hello Web Service Provider!");
             _LIT(msgOK,"Hello World Request: Sent Successfully");
             _LIT(msgFail,"Hello World Request: Failed");
             TInt retVal = iService->SendMessageL( KWSRequestString );
             SetStatus(retVal);
             if (KErrNone == retVal)
             {
                 CCknInfoDialog::RunDlgLD(title, msgOK);
             } else {
                 CCknInfoDialog::RunDlgLD(title, msgFail);
             }
             break;
         }
     default:
         {
             // Abort application if we don't recognize the command
             User::Panic(KWSName, KErrUnknown);
             break;
         }
     }
}
```

Example 10-9. Main client code.

In Example 10-9, the line set in boldface shows a literal text message being sent to the service. The method for sending the actual message (`CHelloWSServiceProc:: SendMessageL`) is as follows:

```
TInt CHelloWSServiceProc::SendMessageL(const TDesC &requestString)
{
    // Create a new hello request message
    TInt retVal = KErrNone;

    CSenElement* helloString = CSenBaseElement::NewL(_L("HelloString"));
    CSenElement* helloRequest =
        CSenBaseElement::NewL(_L("urn:nokia:samples"),
                              _L("RequestHello"));
    CSenSoapMessage* soapRequest = CSenSoapMessage::NewL();

    CSenElement* usernameToken =
        CSenBaseElement::NewL(KSecurityXmlns(),_L("UsernameToken"));
    CSenElement* usernameXML =
        CSenBaseElement::NewL(KSecurityXmlns(), _L("Username"));
    CSenElement* passwordXML =
        CSenBaseElement::NewL(KSecurityXmlns(), _L("Password"));

    CleanupStack::PushL(soapRequest);

    usernameXML->SetContentL( KWSUserName );
    passwordXML->SetContentL( KWSPassword );

    usernameToken->AddElementL( *usernameXML );
    usernameToken->AddElementL( *passwordXML );

    helloString->SetContentL( requestString );
    helloRequest->SetContentL( *(helloString->AsXmlL()) );

    soapRequest->SetSecurityHeaderL( *(usernameToken->AsXmlL()) );
    soapRequest->SetBodyL( *(helloRequest->AsXmlL()) );

    /*
    * You could just simply send a string of XML like this:
    *
    * HBufC8* request = HBufC8::NewLC(256);
    * request->Des().Append(_L8("<RequestHello
    xmlns=\"urn:nokia:samples\"><HelloString>Hello Web Service
    Provider!</HelloString></RequestHello>"));
    * CleanupStack::PopAndDestroy();
    *
    */

    /*
    * Alternatively, you could send an entire SOAP message:
    *
    * retVal = GetConnection()->SendL( *soapRequest );
    *
    * This would send an ASYNCHRONOUS request message to a service.
    * The observing object would deliver the response message via the
    * HandleMessage method specified on the MSenServiceConsumer
    * interface.
    * In addition, the SetStatus callback method would
    * be triggered when the service connection status changes.
    *
    */

    /*
    * The code for the synchronous submission of a SOAP message is as
    follows:
```

```
     *
     */
    HBufC8* helloResponse = NULL;
    TBool aReady = false;
    if (GetConnection()) GetConnection()->IsReady(aReady);
    if (aReady)
    {
        retVal = GetConnection()->SubmitL( *soapRequest, helloResponse);
    } else {
        retVal = KErrSenNotInitialized;
    }
    if ( helloResponse )
    {
        CleanupStack::PushL(helloResponse);
        /*
        * INSERT CODE HERE FOR PROCESSING RESPONSE MESSAGE:
        *
        * HBufC* pUnicode = SenXmlUtils::ToUnicodeLC(helloResponse-
        >Left(200));
        * ...
        * CleanupStack::PopAndDestroy(); // pUnicode
        * ...
        *
        */
        CleanupStack::Pop(helloResponse); // helloResponse
    }

    CleanupStack::PopAndDestroy(soapRequest); // soapRequest

    return retVal;
}
```

Example 10-10. Method for sending a message.

Note that before a message is sent to the Web service provider, a connection to that provider's service has to be established and the service has to be described to the service consumer.

The XML sent in the message is produced by creating objects of the type CSenElement representing those pieces of XML – sort of an XML miniature DOM tree. It is also possible to use only strings, and submit XML strings directly for NWSF to process and send.

10.3 NWSF Service Connection Library

`SenServConn` is a module for handling the establishment of service connections. It includes a callback interface for receiving status, error, and message information.

10.3.1 MSenServiceConsumer Interface

`MSenServiceConsumer` is an interface, which must be implemented by all NWSF client applications. In our example, `CHelloWS` implements the interface in its header file (see example file `HelloWSService.h`) as follows:

```
...
public:
  virtual void ConnectToServiceL();
  virtual TInt SendMessageL(const TDesC &requestString);
  virtual void HandleMessageL(const TDesC8& aMessage);
  virtual void HandleErrorL(const TDesC8& aError);
  virtual void SetStatusL(const TInt aStatus);
...
```

Example 10-11. Implementation of MSenServiceConsumer in CHelloWS header file.

In Example 10-11, four methods are declared as `virtual void`, indicating that they are to be implemented by the class defined in this header file. Note that although the methods required by the `MSenServiceConsumer` interface are defined here, the interface itself is not utilized directly by the implementing class. `MSenServiceConsumer` derives from `CActive` (a Symbian "active object"). If the application is already an active object (and in this case, the HelloWS application is already running in an active object, handling events sent to the user interface), you should not derive the implementation class directly from `MSenServiceConsumer`.

10.3.2 HandleMessage()

The `HandleMessage()` method is called upon the receipt of a message from a remote Web service. A simple implementation is shown in Example 10-12 below.

```
void CHelloWSDocument::HandleMessageL(const TDesC8 &aMessage)
{
    if(aMessage.Length() > 0)
    {
        Log()->Write(_L(" HandleMessageL() received\n"));
        Log()->Write(aMessage);
    }
}
```

Example 10-12. Implementation of HandleMessage.

By default, the messages passed into this method are already stripped of the SOAP envelope and associated XML elements. In other words, the received XML is just the service-specific part of the SOAP message. For Example 10-12, this would appear as follows:

```
<ex:HelloResponse xmlns:ex="urn:nokia:ws:samples">
  <ex:ResponseString>
    Hello Web Service Consumer!
  </ex:ResponseString>
</ex:HelloResponse>
```

Example 10-13. XML data received by HandleMessage.

A client application can ask NWSF to provide it with the complete SOAP-enveloped message by setting a flag on the connection as shown in Example 10-14 below.

```
TBool ON = ETrue ;
iConnection->CompleteServerMessagesOnOff( ON ) ;
```

Example 10-14. Asking NWSF not to remove the SOAP envelope.

Setting the flag to `ON` returns the full SOAP envelope (including the contained service-specific message) to the client application. As some NWSF client applications need to handle more than one message, and need to execute more complex tasks than just log received messages, the `HandleMessage` implementation may actually become an event-handling routine for received messages. In general, the format for a more complex event handler is that the `HandleMessage` implementation first parses the message, and then the client can determine which input message it has received, and based on that, can execute appropriate processing.

```
void CHelloWS::HandleMessageL(const TDesC8& aMessage)
{
    iLog.Write(_L("CHelloWS::HandleMessageL()"));
    CHelloResponse* handler = CHelloResponse::NewL() ;

    RXMLReader* xmlReader = new RXMLReader() ;
    xmlReader->CreateL() ;
    CleanupStack::PushL(xmlReader);
    xmlReader->SetContentHandler(handler);
    xmlReader->ParseL(aMessage);
    CleanupStack::Pop(); //xmlReader
    xmlReader->Destroy();
    if ( handler->Status() == EHello )
    {
        iConsole.Printf(_L(" Hello\n"));
    }
    else if ( handler->Status() == EHowAreYou )
    {
        iConsole.Printf(_L(" How are you?\n"));
    }
    CActiveScheduler::Stop();
}
```

Example 10-15. Creating an XML handler.

In Example 10-15, an XML handler is created for the incoming message, which could be one of two possible input messages (a `HelloResponse`, or a `HowAreYouResponse`). The XML handler is then used to help parse the received XML content. In this case, this XML reader could have been created in the constructor for this class (rather than in `HandleMessage()`), but if you expect to receive multiple messages in a short period of time, you might want to create an XML reader per input message. The handler code for `CHelloResponse` is as follows:

```
#include <badesca.h>
/*
 *
================================================================================
 * Name : HelloResponse.cpp
 *
 * Autogenerated by XSLT translation of WSDL document
 *
================================================================================
 */

#include "CHelloResponse.h"

namespace
{
   _LIT8(KNameTokenFormatString,
"<wsse:UsernameToken><wsse:Username>%S</wsse:Username></wsse:
UsernameToken>");
}

EXPORT_C CHelloResponse* CHelloResponse::NewL()
{
   CHelloResponse* pNew = new (ELeave) CHelloResponse;
   CleanupStack::PushL(pNew);

   pNew->ConstructL();

   CleanupStack::Pop(); // delete pNew;
   return pNew;
}

EXPORT_C CHelloResponse* CHelloResponse::NewL(const TDesC& aData)
{
   CHelloResponse* pNew = new (ELeave) CHelloResponse;
   CleanupStack::PushL(pNew);

   pNew->ConstructL(aData);

   CleanupStack::Pop(); // pNew;
   return pNew;
}

CHelloResponse::CHelloResponse()
{
}

void CHelloResponse::ConstructL()
{
   BaseConstructL(KXmlns, KName, KQName);
}

void CHelloResponse::ConstructL(const TDesC& aData)
{
   BaseConstructL(KXmlns, KName, KQName); AsElement().SetContentL(aData);
}

EXPORT_C CHelloResponse::~CHelloResponse()
{
   delete iCharacters;
}

// End autogenerated code

TInt CHelloResponse::Status()
{
   return iStatus;
```

```
}

TInt CHelloResponse::StartDocument()
{
   return 1;
}

TInt CHelloResponse::EndDocument()
{
   return 1;
}

TInt CHelloResponse::StartElement(TDesC& aUri,
                                  TDesC& aLocalName,
                                  TDesC& aName,
                                  MXMLAttributes* aAttributeList)
{
   _LIT( KHelloResponse, "HelloResponse") ;
   _LIT( KHowAreYouResponse, "HowAreYouResponse") ;
   if ( aLocalName == KHelloResponse )
   {
      iStatus = EHello ;
   }
   else if ( aLocalName == KHowAreYouResponse )
   {
   iStatus = EHowAreYou ;
   }
   return 1;
}

TInt CHelloResponse::EndElement(TDesC& aURI,
                                TDesC& aLocalName,
                                TDesC& aName)
{
   return 1;
}

TInt CHelloResponse::Charecters(TDesC& aBuf,
                                TInt aStart,
                                TInt aLength)
{
   delete iCharacters;
   iCharacters = NULL;
   iCharacters = aBuf.Alloc();
   ReplaceXmlEscaping(iCharacters->Des());
   return 1;
}

TPtrC CHelloResponse::GetLocalName(const TDesC& aQualifiedName) const
{
    TInt index = aQualifiedName.Locate(KNamespaceSep);
    if (index != KErrNotFound && index < aQualifiedName.Length())
    {
        index++; // skip separator
        return aQualifiedName.Mid(index);
    }
    else
    {
        return TPtrC(aQualifiedName);
    }
}

void CHelloResponse::ReplaceXmlEscaping(TDes& aBuf)
{
   CDesCArray* pEscapings = new (ELeave) CDesCArrayFlat(5);
   CDesCArray* pReplacements = new (ELeave) CDesCArrayFlat(5);
   CleanupStack::PushL(pEscapings);
```

```
    CleanupStack::PushL(pReplacements);
    pEscapings->AppendL(KLt);
    pReplacements->AppendL(KLtR);
    pEscapings->AppendL(KGt);
    pReplacements->AppendL(KGtR);
    pEscapings->AppendL(KAmp);
    pReplacements->AppendL(KAmpR);
    pEscapings->AppendL(KApos);
    pReplacements->AppendL(KAposR);
    pEscapings->AppendL(KQuot);
    pReplacements->AppendL(KQuotR);
      for(TInt i = 0;i < pEscapings->MdcaCount();i++)
      {
        TInt pos = 0;
        const TDesC& escaping = pEscapings->MdcaPoint(i);
        TInt length = escaping.Length();
        const TDesC& replacement = pReplacements->MdcaPoint(i);
        do
          {
            pos = aBuf.Find(escaping);
            if(KErrNotFound != pos)
              {
                aBuf.Replace(pos, length, replacement);
              }
          }
        while(KErrNotFound != pos);
      }
    pEscapings->Reset();
    pReplacements->Reset();
    CleanupStack::PopAndDestroy(2); // pReplacements, pEscapings
}
```

Example 10-16. Handler code for CHelloResponse.

The key methods in the above example are the `StartElement`, `EndElement`, `StartDocument`, and `EndDocument` handlers. They are used to enter code that is executed when the named events occur during XML parsing. Thus, when the `StartElement()` method is executed, the status of the response is set by the XML content handler. It is useful to employ a basic content handler such as the one used in Example 10-16 if it is not known which of a number of different messages the application has received. However, if you are expecting to receive a particular message or fragment of XML, you can use the `SenBaseFragment` class (or create your own class inherited from `SenBaseFragment`) to perform some of the content handling.

10.3.3 HandleError()

The `HandleError()` method is structured similarly to `HandleMessage()`. The method is called when the framework detects an error. Some errors result from situations such as the inability to access a mobile network. Other errors might be detected before the sending of a message to a service. Note that from the perspective of NWSF, SOAP faults are considered as valid messages instead of errors, and therefore should be handled with the `HandleMessage()` method.

```
void CHelloWS::HandleErrorL(const TDesC8& aError)
{
   CErrorResponse* handler = CErrorResponse::NewL() ;

   RXMLReader* xmlReader = new RXMLReader() ;
   xmlReader->CreateL() ;
   CleanupStack::PushL(xmlReader);
   xmlReader->SetContentHandler(handler);
   xmlReader->ParseL(aError);
   CleanupStack::Pop(); //xmlReader
   xmlReader->Destroy();

   if ( handler->Status() == EConnectionFailure )
   {
     iConsole.Printf(_L(" No connection available\n"));
   }
   else if ( handler->Status() == EFault )
   {
     iConsole.Printf(_L(" SOAP Fault\n"));
   }

   CActiveScheduler::Stop();
}
```

Example 10-17. HandleError() method.

10.3.4 SetStatus()

The `SetStatus()` callback method is triggered when the status of a connection changes. It can be used to perform processing upon such a change. In the HelloWS example, the active scheduler is stopped when the connection status changes. In the example code, this returns control to the code that constructed the object (which created the connection). That code then sends a message. This processing should be placed within the method itself.

```
void CHelloWS::SetStatus(const TInt aStatus)
{
   iConsole.Printf(_L(" SetStatus(%i) -> "), aStatus);

   switch(aStatus)
     {
     case 0: // connection NEW
       iConsole.Printf(_L(" STATUS: NEW\n"));
       break;
     case 1: // connection READY
       iConsole.Printf(_L(" STATUS: READY\n"));
       break;
     case 2: // connection EXPIRED
       iConsole.Printf(_L(" STATUS: EXPIRED\n"));
       break;
     }
   CActiveScheduler::Stop(); // return control to caller
}
```

Example 10-18. SetStatus() method.

10.4 Sending Messages to a Web Service

As the above examples illustrate, it is easy to establish a connection to a Web service by using NWSF. Once the connection has been established, the application can simply submit messages to the service, either by expecting a synchronous response, or by implementing handlers for the asynchronous interfaces of `MSenServiceConsumer`.

The following line, taken from the HelloWS example, submits a message synchronously to the service: `iConnection->SubmitL( *soapRequest, helloResponse);`

The same message could have been submitted asynchronously with the following line: `iConnection->SendL(*soapRequest);`

In both cases, an object representing the service message is passed to NWSF for transmission to the service. If you do this while using a framework, NWSF adds the SOAP envelope and the possible framework headers. Alternatively, a full SOAP envelope can be submitted to NWSF. However, this means that you have to add all the necessary headers yourself.

10.4.1 Connecting to a Web Service by Using a Framework

By using a framework to access a Web service, an application leaves most of the work of adding SOAP headers to a message to the framework. For example, the Liberty ID-WSF defines SOAP headers for message correlation (so that an application can determine that one message is linked to another through a `messageID` value). This correlation header is added automatically to messages produced using that framework. Frameworks can be plugged into NWSF, which means that it is possible to add new frameworks to handle different ways of accessing a service. Currently, the only one framework (Liberty ID-WSF) has been plugged into NWSF. This framework is listed in the NWSF `SenSessions.xml` configuration file:

```
<Framework xmlns="urn:com.nokia.Sen.idwsf.config.1.0"
class="com.nokia.Sen.idwsf.IdentityBasedWebServicesFramework"/>
```

Example 10-19. Configuring the Liberty ID-WSF framework plug-in.

This framework implements the Liberty Alliance ID-WSF specifications, allowing the developer to simply request for this framework to be used when connecting to a Web service. In this particular case, the connection may require the client application first to authenticate itself to the service via a third party (through an associated IDP) and to acquire the necessary credentials. Frameworks can be specified with a framework identifier – for the Liberty framework, the identifier is ID-WSF. Note that if no framework identifier is specified for a service connection (this is not the same as the NONE specifier described below), the Liberty framework is used by default to connect to the service.

10.4.2 Connecting to a Web Service Without Using a Framework

Basic Web services are ones that do not require any particular framework of SOAP headers or any well-defined policy to allow a requester to access the service. Connecting to these services is also simple, but since the default policy is to use a framework, developers must specifically request NWSF not to use a framework when connecting to a Web service. This is done with an API call `SetFrameworkID(...)`. The following example sets the framework value as `NONE`, which means that the application developer would have to add all necessary SOAP headers for authentication, addressing, and so on to messages before submitting them to NWSF.

```
_LIT(KFrameworkID, "NONE") ;
iDesc->SetFrameworkIdL( (TDesC*) &KFrameworkID()) ;
```

Example 10-20. Deactivating all frameworks.

10.5 NWSF Service Manager Library

The NWSF architecture is administered by the Service Manager (`CsenServiceManager`). This class, provided by the Service Manager library, is the main entry point for NWSF (in addition to the `CSenServiceConnection` class). `CsenServiceManager` is used for service and IDP registration and deregistration, as well as for service association and dissociation. The actual service descriptions, however, are created using Service Description classes.

10.6 NWSF Service Description Library

This library provides a set of classes for managing the description of services, including the creation and manipulation of service descriptions. These descriptions can then be registered into the NWSF service database using NWSF Service Manager.

10.6.1 Using the NWSF Service Database

The NWSF service database is where services can be stored persistently between invocations of NWSF. The Service Manager reads these descriptions when it starts up, and they can then be requested by NWSF client applications.

Service descriptions are created with NWSF Service Description classes and then registered by using NWSF Service Manager. Once the required authentication data has been registered, the service is associated with this information. Based on this information, NWSF uses the selected framework to establish a connection, and if necessary, uses the authentication and discovery services to get the access credentials needed for the actual service.

10.7 NWSF and Identity

Identity is highly important in both Web services and the mobile world. Typically, Web service users (requesters) need to authenticate themselves in order to gain access to the service. The authentication process relies on either the service itself, or some other entity that the service trusts to know the identity of the service requester. Services may, of course, maintain an account for each user. In such situations, the service self-authenticates the user.

To support self-authentication, NWSF supports authentication with WS-Security Username Token or HTTP Basic Authentication. Services may also rely on a separate IDP, which they trust to make assertions regarding the identity of a service-requesting user. That trust is explicitly modeled in NWSF, by associating IDPs with services that trust them to make such assertions.

The Liberty Alliance Identity Web Services Framework (ID-WSF) specifications define a framework for identity-enabled Web services, and NWSF supports the use of this framework in handling identity information.[LIBERTY] In the framework, IDPs are explicitly decoupled from those that expose a service on behalf of an identity, provide service-based on identity-related information, or have no need for identity at all. The Liberty ID-WSF specifications focus on privacy, security, and other related areas. Two of the most useful concepts in the framework are the opaque name identifier, and the `UsageDirective` SOAP header.

10.7.1 Registering an Identity Provider with NWSF

IDPs can be added to NWSF by using an API method call, which appears as follows:

```
_LIT(KPpContract, "urn:nokia:ws:samples");
_LIT(KIdPContract, "urn:liberty:authn:2004-04") ;
_LIT(KIdPEndpoint, "https://example.nokia.com/soap/IdPAS");
_LIT(KIdPProviderID, "https://90.0.0.18") ;
_LIT(KTestAdvisoryAuthnID, "testuser1"); _LIT(KTestPassword, "testuser1");
_LIT(KIdWsfFrameworkID, "ID-WSF");

CSenServiceManager* serviceMgr = CSenServiceManager::NewL() ;

// Creates and registers the IDP as one identity into NWSF database
CSenIdentityProvider* idpDesc = CSenIdentityProvider::NewLC( (TDesC*)
&KIdPEndpoint(),(TDesC*) &KIdPContract() ) ;
idpDesc ->SetProviderID(KIdPProviderID);
idpDesc ->SetUserInfo(KNullDesC, KTestAdvisoryAuthnID, KTestPassword);
serviceMgr->RegisterIdentityProviderL( *idpDesc ) ;

// Creates and registers the actual service that provides that particular
identity
CSenIdentityProvider* authService = CSenIdentityProvider::NewLC((TDesC&)
KIdPEndpoint(), (TDesC&) KIdPContract());
authService->SetProviderID(KIdPProviderID);
authService->SetFrameworkIdL((TDesC*)&KIdWsfFrameworkID());
serviceMgr->RegisterServiceDescriptionL(*authService);
```

Example 10-21. API method call for adding IDPs to NWSF.

A client application, which is used to connect to a specific service using this IDP, might register the provider and associate itself with it when the application is installed.

10.7.2 Associating an Identity Provider with a Service

In order to tell NWSF that a particular service requires the use of a certain IDP (i.e., that the service provider trusts the IDP), this IDP must be associated with the service. NWSF provides a method for this purpose. The method links the IDP identifier with the service identifier (either a specific endpoint, or the contract URI) of the service being associated.

```
serviceMgr->AssociateServiceL(&KPpContract(),&KIdPProviderID() ) ;
```

Example 10-22. Linking a service provider with an IDP.

Above, the IDP identifier (`KIdPProviderID`) is associated with `KPpContract`, either an actual service endpoint (which associates the IDP only with that single service endpoint) or a service contract (allowing one to associate the IDP with all services adhering to a certain schema contract, such as `urn:nokia:ws:samples`).

10.7.3 Using IDP-Based Authentication

In example 10-10, a `HelloRequest` message was submitted to the HelloWS message without using any framework or authentication to the service. To enable the client application to use IDP-based authentication, only few changes need to be made to the original code. In fact, all that needs to be done is to remove the line that sets the framework identifier to `NONE` or replace it with ID-WSF, and change the invocation of the service description constructor to read as follows:

```
_LIT(KSvcContract, "urn:nokia:ws:samples") ;

CSenXmlServiceDescription* serviceDesc =
  CSenXmlServiceDescription::NewLC( KNullDesC(), KSvcContract() ) ;

//serviceDesc->SetFrameworkIdL( (TDesC*) _L("NONE")) ;
```

Example 10-23. Service description constructor for IDP-based authentication.

10.7.4 Using WS-Security Username Token

NWSF provides an implementation of the WS-Security specification, which allows for the authentication of a SOAP message using a security token. One type of security token is the `UsernameToken`, which is used simply to provide a username and password to authenticate the SOAP message. Obviously, this is not the most secure authentication method available, and when sending credentials in plaintext, SSL/TLS should be used to protect the content of the message. This type of authentication is akin to HTTP Basic Authentication, but is done at the SOAP rather than the HTTP layer. The following code excerpt shows how the security token is added to a SOAP message.

```
HBufC8* tokenString = NULL;

CSenWsSecurityHeader::UsernameTokenL( userName, tokenString ) ;
CleanupStack::PushL(tokenString); // push tokenString
CSenWsSecurityHeader* wsseHeader =

CSenWsSecurityHeader::NewLC(*(SenXmlUtils::ToUnicodeLC(*tokenString)));
CleanupStack::PushL( wsseHeader ); // push wsseHeader
// More code
CSenElement* header = wsseHeader->ExtractElement();
if(header)
   {
   soapRequest->AddHeaderL(*header) ;
   }
CleanupStack::PopAndDestroy(2); // wsseHeader, tokenString
```

Example 10-24. Adding a security token to a SOAP message.

10.7.5 Using HTTP Basic Authentication

Some Web services do not use SOAP Message Security (with WS-Security) or third-party authentication via a framework, but still want to authenticate the requester. NWSF supports such Web services, and allows them to use HTTP Basic Authentication with a username and password.

The following code fragment illustrates the authentication process:

```
CSenServiceManager* serviceMgr = CSenServiceManager::NewLC() ;
_LIT(KServiceEndpoint, "http://example.nokia.com/protected/HelloWS");
_LIT(KFrameworkID, "NONE") ;

CSenIdentityProvider* idp =
CSenIdentityProvider::NewLC((TDesC&)KServiceEndpoint() );

_LIT(KUser, "username");
_LIT(KPass, "password");
idp->SetUserInfo(KUser, KUser, KPass);
serviceMgr->RegisterIdentityProviderL(*idp);
iConnection = CSenServiceConnection::NewL(*this, *idp );
```

Example 10-25. Authenticating a requester by using HTTP Basic Authentication.

The Web service endpoint is protected by HTTP Basic Authentication. When the initial request is made, an HTTP 401 error (requesting authentication) is returned to NWSF, which then supplies the credentials as specified above. This message exchange is as follows:

1. Initial request, posting the SOAP message:

```
POST /protected/HelloWS HTTP/1.1
Host: example.nokia.com
Accept: text/xml
Expect: 100-continue
User-Agent: Sen
Content-Length: 623
Content-Type: text/xml
SOAPACTION: ""
<S:Envelope xmlns:S="http://schemas.xmlsoap.org/soap/envelope/">
  <S:Body>
    <RequestHello xmlns="urn:nokia:samples">
      <HelloString>Hello Web Service Provider!</HelloString>
    </RequestHello>
  </S:Body>
</S:Envelope>
```

2. The server responds, stating that the request was unauthorized, and prompting for credentials.

```
HTTP/1.1 401 Unauthorized
WWW-Authenticate: Basic realm="example.nokia.com"
Content-Type: text/xml
Date: Thu, 30 Sep 2004 16:59:15 GMT
Server: Jetty/4.2.21 (Windows 2000/5.0 x86 java/1.4.2_04)
<html>
  <body>Unauthorized!</body>
</html>
```

3. NWSF adds the HTTP Basic Authentication credentials to the request and re-sends.

```
POST /protected/HelloWS HTTP/1.1
Host: example.nokia.com
Accept: text/xml
Authorization: Basic cm9sZTE6dG9tY2F0
Expect: 100-continue
User-Agent: Sen
Content-Length: 623
Content-Type: text/xml
SOAPACTION:
<S:Envelope xmlns:S="http://schemas.xmlsoap.org/soap/envelope/">
  <S:Body>
    <RequestHello xmlns="urn:nokia:samples">
      <HelloString>Hello Web Service Provider!</HelloString>
    </RequestHello>
  </S:Body>
</S:Envelope>
```

4. The server authenticates the request and responds affirmatively.

```
HTTP/1.1 100 Continue
HTTP/1.1 200 OK
Date: Thu, 30 Sep 2004 16:59:17 GMT
Server: Jetty/4.2.21 (Windows 2000/5.0 x86 java/1.4.2_04)
Content-Type: text/xml
Content-Length: 280
<S:Envelope xmlns:S="http://schemas.xmlsoap.org/soap/envelope/">
  <S:Body>
    <ex:HelloResponse xmlns:ex="urn:nokia:ws:samples">
      <ex:ResponseString>
        Hello Web Service Consumer!
      </ex:ResponseString>
    </ex:HelloResponse>
  </S:Body>
</S:Envelope>
```

Example 10-26. Message exchange for authentication request.

10.8 Policy and Services

Services may have different policies regarding the way they authenticate a requester. The policy can vary from requiring no authentication at all to requiring authentication both for the connection from the client (peer entity), and for the actual message(s) sent by the client.

However, a service may also have an entirely different type of policy. In many cases – and particularly when using a framework – the policy is handled by the framework. In the case of the NWSF Liberty ID-WSF framework, the policy regarding the use of various security mechanisms (i.e., which peer-entity and message authentication mechanisms clients should use) is stored by a discovery service, and NWSF itself will obtain such policies there and store them for use by client applications. Client applications themselves can, however, set specific policies for service providers on 1) the IDPs they trust to authenticate a service requester, and 2) whether or not the service demands the use of a particular Internet access point (IAP) from the device (for example, if the service is only available over a VPN IAP).

10.8.1 Registering Service Policy with NWSF

The following code fragment sets a provider policy on use of a specific IAP: `serviceDesc->SetIapIdL( 56152 );` A client application could also supply a list of IDP identifiers that the provider identified in the service description trusts (using the `SetIdentityProviderIdsL(...)` method).

10.9 Configuration

The NWSF database is implemented using two XML files, which are located in the `\epoc32\<Platform>\c\system\data` directory in the Symbian emulator environment. APIs are provided to allow the developer to interact with the configurations. The first of the two files, `SenSessions.xml`, contains the stored service descriptions, and could end up looking as follows (initially it is empty or not present):

```
<SenConfiguration xmlns="urn:com.nokia.Sen.config.1.0">

  <Framework xmlns="urn:com.nokia.Sen.idwsf.config.1.0"
class="com.nokia.Sen.idwsf.IdentityBasedWebServicesFramework"/>

  <ServiceDescription framework="ID-WSF">
    <Contract>urn:liberty:as:2004-04</Contract>
    <Endpoint>http://example.nokia.com/soap/IDPAS</Endpoint>
    <ProviderPolicy/>
    <ProviderID>http://example.nokia.com/soap/</ProviderID>
  </ServiceDescription>

  <ServiceDescription framework="ID-WSF">
    <Contract>urn:liberty:disco:2003-08</Contract>
    <Endpoint>http://example.nokia.com/soap/IDPDS</Endpoint>
    <Credentials> notOnOrAfter="2004-09-28T18:28:44Z" >
      <saml:Assertion xmlns:saml="urn:oasis:names:tc:SAML:1.0: assertion"
       AssertionID="2sxJu9g/vvLG9sAN9bKp/8q0NKU=">
      …
      </saml:Assertion>
    </Credentials>
    <ProviderPolicy/>
    <ProviderID>http://example.nokia.com/soap/</ProviderID>
    <ResourceID>http://example.nokia.com/soap/IDPDS/4b3d</ResourceID>
    <TrustAnchor>http://example.nokia.com/soap/</TrustAnchor>
  </ServiceDescription>

  <ServiceDescription framework="ID-WSF">
    <Contract>urn:nokia:ws:samples</Contract>
    <Endpoint>http://example.nokia.com/soap/HelloWS</Endpoint>
    <Credentials notOnOrAfter="2004-09-28T18:28:44Z">
      <saml:Assertion xmlns:saml="urn:oasis:names:tc:SAML:1.0:assertion"
       AssertionID="2sxJu9g/vvLG9sAN9bKp/8q0NKU=">
```

```
    ...
      </saml:Assertion>
    </Credentials>
    <ProviderPolicy/>
    <ProviderID>http://example.nokia.com/soap/</ProviderID>
    <ResourceID>http://example.nokia.com/soap/HelloWS/5678</ResourceID>
    <TrustAnchor>http://example.nokia.com/soap/</TrustAnchor>
  </ServiceDescription>

</SenConfiguration>
```

Example 10-27. SenSessions.xml.

The configuration in Example 10-27 defines three services, as well as a framework (Liberty ID-WSF). The services are an IDP, a discovery service, and an actual application service (a more advanced version of the HelloWS service). Authentication for the service is provided by a third-party IDP, which supplies credentials to the service requester. The file `SenIdentities.xml` contains information about the IDPs that are associated with users. Similar to the `SenSessions.xml` configuration file, this file becomes populated by classes of the `SenServDesc` library. An example of this file is presented in Example 10-28 below.

```
<Users>
  <User>
    <Identity>
      <IdentityProvider>
        <Endpoint>http://selma.ndhub.net:9080/tfs/IDPSSO_IDWSF</Endpoint>
        <Contract>urn:liberty:as:2004-04</Contract>
      </IdentityProvider>
    </Identity>
  </User>
</Users>
```

Example 10-28. SenIdentities.xml.

The `SenIdentities.xml` file is used internally by NWSF, primarily for caching service registration information in order to optimize server lookup and authentication processing. Consequently, access to the file is restricted to the APIs defined for registering and associating services.

10.10 NWSF Utility Library

The Service Development API contains the following utility classes:

- SenDateUtils
- CSenGuidGen
- CSenSoapEnvelope
- CSenSoapFault

- CSenSoapMessage
- CSenWsSecurityHeader

The SOAP helper classes are used to support SOAP message construction with WS Security header information (`CSenWsSecurityHeader`), to set the SOAPAction header, and to process SOAP Faults (`CSenSoapFault`). In addition, there are utility classes for converting XML Schema's dateTime datatype to and from a Symbian TTime object (`SenDateUtils`), and for generating unique identifiers (`CSenGuidGen`).

10.11 Description of wscexample Client Code

The wscexample application described in this section approaches the services described elsewhere in this chapter from a new perspective. It combines the features of HelloWS and PhonebookEx (covered below in section 10.13), with a small exception: instead of using the IDPDS and IdPAS demo servlets, it uses Forum Nokia's AddressBook service. It is also different in another way when compared with the PhonebookEx application: the wscexample is a console-based application, while PhonebookEx has a graphical user interface.

Although the application development part of this book concentrates on the Series 80, wscexample is thoroughly commented to also cover the SOA for S60 platform APIs. To port the application to the SOA for S60 platform, uncomment the lines containing SOA for S60 platform code, and comment the respective Series 80 API code lines.

As all relevant NWSF APIs have already been described in the HelloWS example, that information is not repeated here. Instead, we show four key sections of the application, which are used to complete the following tasks:

1. Register a username and a password for HTTP Basic Authentication
2. Establish a connection to the service
3. Create a callback function for reacting to different phases of the service message exchange
4. Create a basic query by using XML elements, and send it using a SubmitL (SOAP) API call

Note that you need to modify the endpoints according to the network topology of your test network.

10.11.1 Registering Credentials

The following code segment demonstrates how to register a username and a password for accessing services secured with HTTP Basic Authentication.

```
void CWscExampleService::ConstructL(CWscExampleObserver& aObserver,
                                    TDesC& aServiceId,
                                    TDesC& aUser,
                                    TDesC& aPwd,
                                    const TDesC* aFrameworkId)
    {
    CSenServiceManager* serviceMgr = CSenServiceManager::NewLC();
    CSenIdentityProvider* idpServiceDesc = NULL;
    TInt retVal(KErrNone);

    if ( *aFrameworkId == KWSFrameworkNone
         || *aFrameworkId == KWSFrameworkBasic )
        {
        // aServiceId is the actual service endpoint
        idpServiceDesc = CSenIdentityProvider::NewLC( aServiceId );

        idpServiceDesc->SetUserInfo(aUser, aUser, aPwd);

        //idpServiceDesc->SetFrameworkIdL(*aFrameworkId); // S60 API
        idpServiceDesc->SetFrameworkIdL((TDesC*)aFrameworkId); // Series 80
API

        // If BASIC AUTH credentials are used by a basic Web service,
        // you need to register user information as follows:
        retVal = serviceMgr->RegisterIdentityProviderL(*idpServiceDesc);

        // NOTE: If BASIC AUTH is not used, you could simply create
        // a service connection by creating a service description containing
        // only an endpoint and a basic ws framework id. This version
        // does not require the Service Manager API registration call.

        CleanupStack::PopAndDestroy(); // idpServiceDesc
        }
```

Example 10-29. Registering a username and a password for HTTP Basic Authentication.

10.11.2 Establishing a Service Connection

The following snippet shows how to establish a service connection by using only endpoint information.

```
        void CWscExampleService::ConstructL:

        // In this case, aServiceId is actually an endpoint to a basic Web
service
        iDesc = CSenXmlServiceDescription::NewL((TDesC&)aServiceId,
KNullDesC); // Series 80 API
        iDesc->SetFrameworkIdL((TDesC*)aFrameworkId);    // Series 80 API
        iConnection = CSenServiceConnection::NewL(iConsumer, *iDesc);
          // Now, wait for SetStatus() callback before attempting to send
the message
```

Example 10-30. Using endpoint information to create a service connection.

10.11.3 Handling Callbacks

The callback code snippet below is used to handle different events raised during the sample application's lifetime. The events handle issues such as the establishment of connections and credential-related matters.

```
void CWscExampleObserver::SetStatus(const TInt aStatus)
    {

    HBufC8* pResponse = NULL;
    TInt retVal(KErrNone);
    TInt leaveCode(KErrNone);

    iConsole.Printf( _L("CWscExampleObserver::SetStatus(%d)"), aStatus );


    switch ( aStatus )
        {
        case KSenConnectionStatusNew: // connection NEW
            {
            iConsole.Printf(_L(" - NEW: KSenConnectionStatusNew\n"));
            }
            break;

        case KSenConnectionStatusReady: // connection READY
            {
            iInitialized = ETrue;
            iConsole.Printf(_L(" - READY: KSenConnectionStatusReady\n"));

            //
            // Send the request message to the WSP
            //
            CSenXmlServiceDescription* initializer =
                iService->ConnectionInitializer();

            if(initializer)
                {
                TPtrC frameworkID = initializer->FrameworkId();
                iConsole.Printf(_L("- Framework:\n    '%S'\n"),
&frameworkID);

                // Invoke the WS-I service
                iConsole.Printf(_L(" - Invoking WS-I service\n"));
                TRAP(leaveCode, retVal =
                     iService->SendExampleQueryL( KWscExampleQueryString,
pResponse ); )

                // Handle leave codes
                if(leaveCode!=KErrNone)
                    {
                    iConsole.Printf( _L("\n - Send query leaved: %d\n"),
leaveCode );

                    if(retVal==KErrNone)
                        {
                        retVal = leaveCode; // return leave code instead
                        }
                    }

                CleanupStack::PushL(pResponse);

                //
                // Handle errors of pResponse
```

```
                //
                switch ( retVal )
                    {
                    case KErrNone:
                        // Success
                        iConsole.Write(_L("OK - response received."));

                            /*
                            * INSERT CODE HERE, FOR PROCESSING RESPONSE
MESSAGE:
                            */
                        break;

                    //Service Development API Errors
                    case KErrSenServiceConnectionBusy:
                        iConsole.Write(_L("KErrSenServiceConnectionBusy\
n"));

                        break;
                    case KErrConnectionInitializing:
                        iConsole.Write(_L("KErrConnectionInitializing\n"));
                        break;
                    case KErrConnectionExpired:
                        iConsole.Write(_L("KErrConnectionExpired\n"));
                        break;
                    case KSenConnectionStatusReady:
                        iConsole.Write(_L("KSenConnectionStatusReady\n"));
                        break;
                    case KSenConnectionStatusExpired:
                        iConsole.Write(_L("KSenConnectionStatusExpired\n"));
                        break;
                    case KErrSenNotInitialized:
                        iConsole.Write(_L("KErrSenNotInitialized\n"));
                        break;

                    default:
                        {
                        iConsole.Printf( _L("\n - Error code: %d\n"), retVal
);

                        break;
                        }
                    } // switch ( retVal )

                // print out the response
                if(pResponse)
                    {
                    iConsole.Printf(_L("\n - Response: (%d bytes): '%S'\n"),
                            pResponse->Length(),
                            SenXmlUtils::ToUnicodeLC(pResponse->Left(134)));

                    CleanupStack::PopAndDestroy(); // ToUnicodeLC

                    }
                CleanupStack::PopAndDestroy(); // pResponse
                } // if(initializer)

            } // case KSenConnectionStatusReady
            break;

        case KSenConnectionStatusExpired: // credentials for this connection
are EXPIRED
            {
            iConsole.Printf(_L("\n - Credentials are expired: KSenConnection
StatusExpired\n"));
            }
            break;
```

```
    }
    iConsole.Printf(_L("- Query submitted and processed.\n"));
    iConsole.Printf(_L("- PRESS ANY KEY TO CLOSE THIS TEST -\n"));
```

Example 10-31. Handling callback events.

10.11.4 Creating a Query

The fourth and final code snippet handles the actual SOAP message. The code example creates the necessary XML-based content, including the security token, and sends the complete message using a synchronous method.

```
TInt CWscExampleService::SendExampleQueryL(const TDesC& aRequest, HBufC8*&
aResult) // Series 80 API
    {

    CSenSoapMessage* soapRequest = CSenSoapMessage::NewL();
    CleanupStack::PushL(soapRequest);

    // Form the username token for the SOAP request:
    CSenElement* usernameTokenElement =
        CSenBaseElement::NewL(KSecurityXmlNsFixed(), KUsernameToken);
    CleanupStack::PushL(usernameTokenElement);

    CSenElement* usernameElement =
        CSenBaseElement::NewL(KSecurityXmlNsFixed(), KUsername);
    CleanupStack::PushL(usernameElement);
    // Set the username as content of this element:
    usernameElement->SetContentL( KBasicUsername );

    // AddElementL call transfers the ownership of
    // usernameElement to usernameTokenElement:
    usernameTokenElement->AddElementL( *usernameElement );
    CleanupStack::Pop(); // usernameElement

    CSenElement* passwordElement =
        CSenBaseElement::NewL(KSecurityXmlNsFixed(), KPwd);
    CleanupStack::PushL(passwordElement);
    // Set the password as content of this element:
    passwordElement->SetContentL( KBasicPassword );
    // AddElementL call transfers the ownership of
    // passwordElement to usernameTokenElement:
    usernameTokenElement->AddElementL( *passwordElement );
    CleanupStack::Pop(); // passwordElement

    //HBufC8* tokenAsXml = usernameTokenElement->AsXmlL();    // S60 API
    HBufC* tokenAsXml= usernameTokenElement->AsXmlL();        // Series 80
API
    CleanupStack::PushL(tokenAsXml);

    soapRequest->SetSecurityHeaderL( *tokenAsXml );

    CleanupStack::PopAndDestroy(2); // tokenAsXml, usernameTokenElement


    // Form the request from XML elements:
    CSenElement* requestElement =
        CSenBaseElement::NewL( KSamplesNamespace, KRequest);
    CleanupStack::PushL(requestElement);
```

```
    CSenElement* dataElement = CSenBaseElement::NewL( KData );
    CleanupStack::PushL(dataElement);
    dataElement->SetContentL( aRequest );
    // AddElementL call transfers the ownership of dataElement to
requestElement:
    requestElement->AddElementL(*dataElement);
    CleanupStack::Pop(); // dataElement

    //HBufC8* requestAsXml = requestElement->AsXmlL(); // Series80 API
    HBufC* requestAsXml = requestElement->AsXmlL(); // Series80 API
    CleanupStack::PushL(requestAsXml);

    soapRequest->SetBodyL( *requestAsXml );
    CleanupStack::PopAndDestroy(2); // requestAsXml, requestElement

    //
    // The following code performs a synchronous submission of a single SOAP
message:
    //
    TBool ready(EFalse);
    if ( iConnection ) iConnection->IsReady(ready);

    TInt retVal(KErrNone);
    if ( ready )
        {
        retVal = iConnection->SubmitL( *soapRequest, aResult);
        }
    else
        {
        retVal = KErrSenNotInitialized;
        }

    CleanupStack::PopAndDestroy(); // soapRequest

    return retVal;
    }
```

Example 10-32. Creating and sending a query.

This concludes the glance at the wscexample application. The complete application is available for download from the Web sites of Forum Nokia and Wiley. We encourage you to study the example code, because the comments provide insight into the functionality of the applications, as well as into the differences when compared with the SOA for S60 platform version.

10.12 Test Service Code

The wscexample client code uses simple services that are based on Java servlets to help test the example. As the servlets are identical to those described in chapter 9 discussing Java client development, please refer to section 9.8.1 for more information.

Note that to demonstrate its ID-WSF capabilities, the wscexample application relies on the public AddressBook service provided by Forum Nokia. Thus, the dummy Liberty servlets are not required in order to test this client.

10.13 PhonebookEx Example Application

We conclude this chapter by introducing the C++ version of the PhonebookEx application, which accompanies the Series 80 Developer Platform SDK. The example application is a Liberty ID-WSF–compliant Web service client, which can be used to search for data from an address book service (Figure 10-2).

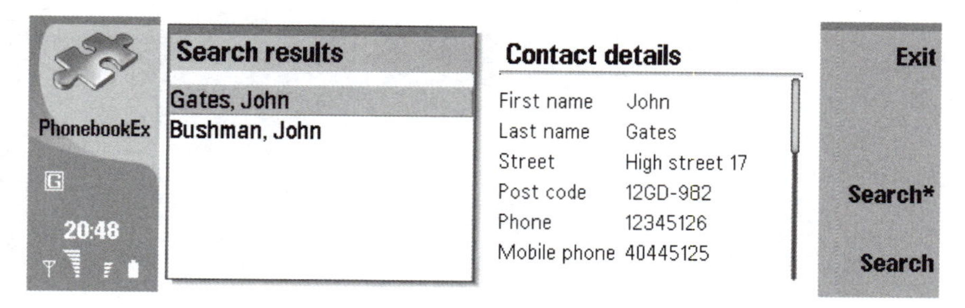

Figure 10-2. PhonebookEx main user interface.

The application UI consists of two search dialogs used to realize a wildcard search or an exact search. The main screen is divided into two sections: on the left is a list of the persons who match the search terms, and on the right is detailed information about the currently selected person. Command buttons are used to launch the search functions and to quit the program.

The following two sections illustrate how to test the PhonebookEx application. The first section provides instructions on how to build the application by using Microsoft Visual Studio .Net 2003, which allows even non-Symbian C++ application developers to test the application. The second section provides a brief discussion on key classes enabling the Web service features.

10.13.1 Building the Application

The example is located in \Symbian\7.0s\S80_DP2_0_PP_SDK\Series80Ex\PhonebookEx (the location of the Symbian folder depends on your installation).

To build the application, test it on an emulator, and deploy it on a mobile device, follow these steps:

1 Start Windows CMD.exe.

2 Go to **\Symbian\7.0s\S80_DP2_0_PP_SDK\Series80Ex\PhonebookEx\group**.

3 To create make files, **run bldmake bldfiles**.

4 To create a MS Visual Studio .Net 2003 project, run **abld makefile vc7**.

5 Go to **\Symbian\7.0s\S80_DP2_0_PP_SDK\epoc32\BUILD\SYMBIAN\7.0S\S80_DP2_0_PP_SDK\SERIES80EX\PHONEBOOKEX\GROUP\PHONEBOOKEX\WINS** and build a project using MS Visual Studio .Net 2003.

6 Open file **AddressBookEngine.cpp** and comment out line 272: //iObserver.ErrorL(ETooManyMatches); The application cannot handle this exception.

7 To configure the emulator network settings, run **\Symbian\7.0s\S80_DP2_0_SDK\Epoc32\Tools\network_config.bat**.

8 To start the emulator, run **\Symbian\7.0s\S80_DP2_0_PP_SDK\Epoc32\Release\Wins\Udeb\Epoc.exe**.

9 Ensure that the emulator network connection is fully functional by testing it with the emulator's Web browser.

10 Start PhoneBookEx, select **search*** and type **v**, or select **search** and type **John**.

11 Go to **\Symbian\7.0s\S80_DP2_0_PP_SDK\Series80Ex\PhonebookEx\group**.

12 Run **abld build thumb urel**.

13 **Go to Symbian\7.0s\S80_DP2_0_PP_SDK\Series80Ex\PhonebookEx\sis**.

14 To create a SIS package for a mobile device, run **makesis PhonebookEx.pkg**.

15 Connect your device to Nokia PC Suite with a USB cable.

16 Start Nokia PC Suite, and select **Install Application**, then select **NokiaWSF.sis**, and install it first.

17 Install **PhonebookEx.SIS**.

18 Test the application on your Series 80 device.

10.13.2 PhonebookEx Web Service Consumer

The PhonebookEx application consists of ten separate files. This section focuses only on the file that handles Web service–related tasks.

The Web service operations of the PhoneBookEx project reside in `AddressBookEngine.cpp` file. The class inherits from `CSenBaseFragment` and `MSenServiceConsumer` classes. `CSenBaseFragment` is a base fragment class equipped with parsing functionalities. `MSenServiceConsumer` is an interface class for `SenServiceConnection` callbacks made from remote service. All service-consuming applications must implement this interface.

The following list describes the connection initialization process. Please refer to the respective numbers in the code snippet below:

1. The connection parameters are the same as in the Java-based AddressBook example (see Table 9.1).
2. `CSenServiceManager` is the main entry point for NWSF along with the `CSenServiceConnection` class. The class is used for service and IDP registration and deregistration, as well as for service association and dissociation.
3. `CSenIdentityProvider` class is used for identity registration with CSenServiceManager. It inherits the `CSenXmlServiceDescription` class and adds user information management (mainly username/password) into it.
4. `RegisterIdentityProviderL()` is used by applications to inform the Service Manager about a service (for example, an authentication service, which typically cannot be discovered through other means).
5. `RegisterServiceDescriptionL()` is used by applications to inform the Service Manager about a service (for example, an authentication service, which typically cannot be discovered through other means).
6. `AssociateServiceL()` associates a service with an IDP. It allows an application to define for the service the IDP with which it wants to interact.

```
File: AddressBookEngine.cpp
/*****************************************************************
InitializeServicesL() initializes the services to NWSF. It creates the
CSenServiceManager instance and uses it to register
    IdentityProvider and ServiceDescription. The service is also
    associated to the registered IdentityProvider for further use.
*****************************************************************/
TInt CAddressBookEngine::InitializeServicesL()
    {
```

```
    _LIT(KPpContract, "urn:nokia:test:addrbook:2004-09");//  1.
    _LIT(KASEndPoint, "http://selma.ndhub.net:9080/tfs/IDPSSO_IDWSF");
    _LIT(KASContract, "urn:liberty:as:2004-04");
    _LIT(KASProviderID, "http://selma.ndhub.net:9080/tfs/");
    _LIT(KTestAdvisoryAuthnID, "testuser1");
    _LIT(KTestPassword, "testuser1");
    _LIT(KIdWsfFrameworkID, "ID-WSF");

    TInt retVal(KErrNone);

    if(!iManager)
        iManager = CSenServiceManager::NewL(); // 2.

    CSenIdentityProvider* identityPrvdr = // 3. CSenIdentityProvider::NewLC(
        (TDesC&)KASEndPoint(), (TDesC&)KASContract());

    identityPrvdr->SetProviderID(KASProviderID);
    identityPrvdr->SetUserInfo(KNullDesC, KTestAdvisoryAuthnID,
KTestPassword);

retVal = iManager-> RegisterIdentityProviderL(*identityPrvdr);// 4.
    if(retVal != KErrNone && retVal != KErrAlreadyExists)
        {
        CleanupStack::PopAndDestroy(); // identityPrvdr
        return -1; // abort
        }

    CSenIdentityProvider* authService = CSenIdentityProvider::NewLC(
        (TDesC&)KASEndPoint(), (TDesC&)KASContract());
    authService->SetProviderID(KASProviderID);
    authService->SetFrameworkIdL((TDesC*)&KIdWsfFrameworkID());

retVal = iManager-> RegisterServiceDescriptionL(*authService);// 5.
    if(retVal != KErrNone && retVal != KErrAlreadyExists)
        {
        CleanupStack::PopAndDestroy(2); // authService, identityPrvdr
        return -1; // abort
        }

retVal = iManager->AssociateServiceL(KPpContract(), // 6. KASProviderID());

    if(retVal != KErrNone && retVal != KErrNotFound) //
        {
        CleanupStack::PopAndDestroy(2); // authService, identityPrvdr
        return -1; // abort
        }

    CleanupStack::PopAndDestroy(2); // authService, identityPrvdr

    return retVal;
    }
```

Example 10-33. Code for PhonebookEx example.

Appendix A: Web Services Standards Organizations

In the preceding chapters, we have introduced several technologies and applications of these technologies. In this appendix, we present the four major standards organizations behind the specifications referred to in this book. The organizations continue to have a significant impact on the evolution of Web services.

A.1 W3C

The World Wide Web Consortium (W3C) has the stated goal of creating universal access to the Web, a Semantic Web, and a Web of Trust. This means that not only is the W3C about the Web, but also about a Web that allows machine interaction and processing, and includes security mechanisms to enable trust in different parties and information on the Web. According to the W3C, its role is to provide vision, design and standardization leadership to achieve its goals. To do this, it applies the key design principles of interoperability, evolution, and decentralization. (Source: W3C, "About the World Wide Web Consortium")

A.1.1 Value Proposition and Membership Base

There is great value in the W3C vision and architectural and design leadership, as well as in its process that supports vendor neutrality, coordination across organizations, and the use of consensus as much as possible. The work of the W3C, ranging from the World Wide Web (HTML), data modeling (XML specification family), Web Services (SOAP, WSDL), security (XML Signature and XML Encryption) to the Semantic Web and the Resource Description Framework (RDF), has had a far-reaching impact on companies, industries and other standards organizations. Other organizations that have used the work of the W3C include Organization for the Advancement of Structured Information Standards (OASIS), the Liberty Alliance, and the Web Services Interoperability Organization (WS-I). For more information on the W3C, visit <http://www.w3.org/>.

A.1.2 Organization and Operating Model

The W3C is a member-driven consortium. Member organizations participate in working groups to produce consortium recommendations and other deliverables. Each member company also has an Advisory Committee Representative (AC-Rep), who can attend two Advisory Committee meetings each year to receive updates on the W3C status. The W3C has over 350 member organizations.

"W3C Activities are generally organized into groups: Working Groups (for technical developments), Interest Groups (for more general work), and Coordination Groups (for communication among related groups). These groups are made up of representatives from Member organizations, the Team, and invited experts. The groups produce the bulk of W3C's results: technical reports, open source software, and services (e.g., validation services). These groups also ensure coordination with other standards bodies and technical communities. There are currently over fifty W3C Working Groups." (Source: W3C, "About the World Wide Web Consortium") W3C activities are led and managed by the W3C Team, which consists of paid team members, fellows from member organizations, and unpaid volunteers. The W3C has approximately sixty team members based in three international locations: MIT/LCS in the United States, ERCIM headquarters in France, and Keio University in Japan. The W3C team is led by the Chief Operating Officer (Steve Bratt) and the Director (Tim Berners-Lee). For more information on W3C members, visit <http://www.w3.org/People/>.

The W3C Team receives advice about the management and operation of the consortium from an Advisory Board, which, unlike a board of directors, does not make binding decisions but rather gives advice. The Advisory Board has a chair appointed by the Team, but to date this has been the W3C Chair. The Advisory Board also has nine elected members with staggered two-year terms. For more information on the W3C Advisory Board, visit <http://www.w3.org/2002/ab/>.

The Technical Architecture Group (TAG) focuses on technical issues in keeping with its mission of providing "stewardship of the Web architecture." The TAG has five elected members with staggered two-year terms. The appointed chair is Tim Berners-Lee (W3C Director). For more information on the W3C Technical Architecture Group, visit <http://www.w3.org/2001/tag/>.

To facilitate management, the W3C Team organizes W3C Activities and other work into three domains and one initiative:

1. The Architecture Domain develops core technologies.
2. The Interaction Domain seeks to improve user interaction, facilitate single Web authoring, and develop formats and languages to present information to users with accuracy, beauty, and a higher level of control.
3. The Technology and Society Domain seeks to develop the Web infrastructure to address social, legal, and public policy concerns.
4. The Web Accessibility Initiative (WAI) promotes usability for people with disabilities through five primary areas of work: technology, guidelines, tools, education and outreach, and research and development.

A.1.3 Deliverables

Once they have been approved by the Director, core technology specifications developed in the W3C become W3C Recommendations (the W3C equivalent of an approved standard). These Recommendations form the basis for implementations and further technology development in the W3C or other organizations. For example, the XML Signature Syntax and Processing Recommendation specifies how digital signature technology can be used in XML technologies. This Recommendation forms the basis of the OASIS SOAP Message Security standard, as well as many other standards ranging from Digital Rights Management to e-business standards (e.g., ebXML messaging). The W3C has also provided sample implementations, such as software libraries, an Open Source browser (Amaya) and a Web server (Jigsaw), and tools (e.g., Tidy for HTML validation) to help disseminate knowledge and foster technology adoption.

The W3C also does much in the area of outreach, from International World Wide Web conferences to workshops and education on the creation of accessible Web sites.

The detailed list of W3C deliverables is far too long to be listed here (it is available on the W3C Web site). Nonetheless, some important deliverables are mentioned in the next section.

A.1.4 Achievements

The following is a brief summary of some major W3C achievements, including key Recommendations. For more publications and details, see the W3C Technical Reports and Publications.

- **Q4 1994** – W3C Founded
- **Q2 1996** – Public release of Jigsaw, the modern object-oriented Web server
- **Q3 1996** – Portable Network Graphics (PNG) Specification (second edition), (revised Q3 2003)
- **Q3 1996** – Platform for Internet Content Selection (PICS) Recommendation
- **Q4 1996** – Cascading Style Sheets (CSS) Level 1 Specification, (revised Q1 1999)
- **Q1 1997** – HyperText Markup Language (HTML) 3.2 Reference Specification
- **Q1 1997** – Public release of Amaya, W3C Editor and Browser
- **Q1 1998** – Advisory Board established
- **Q1 1998** – Extensible Markup Language (XML) 1.0 (Third Edition), (revised Q1 2004)
- **Q4 1999** – HTML 4.01
- **Q4 1999** – Revision of W3C Patent Policy begins
- **Q1 2000** – Extensible HyperText Markup Language (XHTML) 1.0 (second edition)
- **Q1 2001** – Technical Architecture Group (TAG) established
- **Q2 2001** – XML Schema Part 0-2 Recommendations

- **Q2 2001** – XHTML 1.1 (module-based XHTML)
- **Q3 2001** – XML Information Set (second edition)
- **Q3 2001** – Scalable Vector Graphics (SVG) 1.1 Specification (revised Q1 2003)
- **Q3 2001** – Public review of W3C Patent Policy Framework begins
- **Q4 2001** – Extensible Stylesheet Language (XSL) 1.0
- **Q1 2002** – XML-Signature Syntax and Processing Recommendation
- **Q2 2002** – Platform for Privacy Preferences (P3P 1.0)
- **Q3 2002** – Exclusive XML Canonicalization
- **Q4 2002** – XML Encryption Syntax and Processing Recommendation
- **Q1 2003** – DOM Level 2
- **Q1 2003** – XPointer Recommendations
- **Q2 2003** – SOAP 1.2 Recommendations
- **Q2 2003** – Revised W3C Patent Policy approved (see References)
- **Q4 2003** – XForms 1.0
- **Q1 2004** – 5 Feb 2004 Patent Policy effective (see W3C, Patent Policy)
- **Q1 2004** – XML Information Set (second edition, revised)
- **Q1 2004** – Namespaces in XML 1.1
- **Q1 2004** – Resource Description Framework (RDF) Recommendations
- **Q1 2004** – OWL Web Ontology Recommendations
- **Q2 2004** – DOM Level 3 Recommendations
- **Q2 2004** – Tim Berners-Lee receives the inaugural Millennium Technology Prize of one million euros for inventing the World Wide Web
- **Q3 2004** – Tim Berners-Lee is made Knight Commander, Order of the British Empire, by Queen Elizabeth
- **Q4 2004** – Architecture of the World Wide Web, Volume One Recommendation

- **Q1 2005** – Attachments: SOAP Message Transmission Optimization Mechanism, XML-binary Optimized Packaging, Resource Representation SOAP Header Block
- **Q2 2005** – XML Key Management v2.0 (XKMS)

A.1.5 Influence and Future Directions

The W3C has had an enormous influence on the development and evolution of the World Wide Web, XML technologies, and Web Services. This influence has included architecture, standards creation, and technology transfer. The technology transfer has included W3C open source software, such as the Amaya browser, the Jigsaw Web server, and tools such as Tidy (see W3C, Open Source Software). The W3C has also done much in the area of building consensus and market awareness, for example, in the form of the International World Wide Web conferences, workshops, and other activities.

Other organizations continue to build on W3C recommendations. The W3C is driving a vision of the future with respect to accessibility, the Semantic Web, and the Web of Trust. W3C is also exploring issues related to mobility (see W3C, Device Independence).

A.2 OASIS

The Organization for the Advancement of Structured Information Standards (OASIS) is an international not-for-profit consortium devoted to the development, convergence, and adoption of e-business standards. It was founded as SGML Open in 1993, when it focused on the Standard Generalized Markup Language (SGML). In 1998, the organization broadened its scope to include XML and related standards, and changed its name to OASIS. The organization currently supports a wide range of activities, including important standardization work in the area of Web Services, security, and e-business. Although a major aspect of the organization's work is the creation of technical specifications, it is also driving the adoption of standards through marketing and other means. OASIS committees span a range from horizontal technical standards efforts to vertical industry standards and marketing and adoption committees. For more information on OASIS, visit <http://www.oasis-open.org/>.

A.2.1 Value Proposition and Membership Base

OASIS has over 4,000 participants representing more than 600 organizations and individual members in 100 countries. One of the strengths of OASIS is that it is an open organization that allows members to initiate work that they consider to be important. The organization does not restrict its activities, as long as they are within the scope of the OASIS mission, follow the general policies of the organization, and are supported by an adequate number of members. This approach allows a variety of activities to be initiated and strong results to emerge.

OASIS is distinguished by its transparent governance and operating procedures. Members themselves set the OASIS technical agenda, using an open process expressly designed to promote industry consensus and unite disparate efforts. Completed work is ratified by open ballot. Officers of both the OASIS Board of Directors and Technical Advisory Board are chosen by democratic election to serve two-year terms. Consortium leadership is based on individual merit and is not tied to financial contribution, corporate standing, or special appointment. (OASIS, About OASIS)

A.2.2 Organization and Operating Model

OASIS is a member-driven organization. Members initiate technical committees and participate in them to produce results, such as technical standards. The consortium supports this activity by providing policies and processes to enable results to be achieved in an effective and practical manner. The OASIS staff, led by Patrick Gannon (President and CEO), manages the organization, maintains communications to ensure industry awareness of OASIS work, leads the standardization process, and manages OASIS membership. For more information on the OASIS staff, visit <http://www.oasis-open.org/who/staff.php>. The Board of Directors is responsible for the overall governance of OASIS. Except for the OASIS President, who is a permanent member of the Board, the Board members are elected as individuals from the OASIS sponsor- and contributor-level membership organizations to serve two-year terms. Information on the OASIS board is available at <http://www.oasis-open.org/who/bod.php>.

The Technical Advisory Board (TAB) provides guidance to OASIS on issues related to strategy, process, interoperability, and scope of OASIS technical work. Information on the OASIS TAB is available at <http://www.oasis-open.org/who/tab.php>. The primary mechanism for OASIS work is the Technical Committee (TC). A TC is initiated by OASIS members, and works according to the scope of its charter (which is established when the TC is initiated). OASIS currently has over 80 TCs (as well as numerous sub-committees). The TCs are known for producing technical standards that are widely adopted in the industry. Well-known examples include SOAP Message Security (WS-Security 2004) and the Security Assertion Markup Language (SAML) produced by the Web Services Security (WSS) TC and the Security Services TC (SSTC), respectively. The work of TCs has public visibility through publicly available mail archives. OASIS members can influence specifications by joining technical committees, by participating in the public reviews preceding standardization, and by voting on standardization (if they are voting members).

OASIS also has a number of active Member Sections, which are focus areas that may include several technical committees. Many member sections were formerly independent standards organizations that have been incorporated into OASIS. The currently active OASIS Member Sections include CGM Open, DCML, LegalXML, PKI, and UDDI.

In addition, OASIS provides industry Information Channels, including the widely respected Cover Pages <http://xml.coverpages.org/>, XML.org <http://www.xml.org/>, and ebXML.org <http://www.ebxml.org/>.

A.2.3 Deliverables

The primary deliverables of OASIS have been standards related to XML and e-business. The organization has produced a number of important standards that are applicable to many industries. These standards include, for example, Web Services Security (SOAP Message Security), Security Assertions Markup Language (SAML), XACML, and UDDI. OASIS has produced significant e-business standards, such as ebXML and the Universal Business Language (UBL).

OASIS also has numerous technical committees working on topics appropriate to specific industry verticals, such as industry-specific vocabularies. For example, the legalXML member section is active in applying XML technology to the needs of the legal community.

Finally, OASIS works toward the adoption of technologies by hosting events such as educational sessions, symposia, standards days, and interoperability test events.

A.2.4 Achievements

OASIS has a strong influence on horizontal and vertical technologies related to XML. OASIS produces technical standards as one of its deliverables. The following list contains a selection of OASIS standards released to date, as well as other milestones:

- **1993** – SGML Open founded
- **1995-1999** – SGML Table model standards
- **1996** – SGML Entity management standard
- **1996** – SGML Fragment interchange standard
- **1998** – SGML Open renamed OASIS
- **Q1 2001** – Docbook 4.1 standard
- **Q1 2002** – Directory Services Markup Language (DSML) 2.0
- **Q1 2002** – ebXML Registry Services Specification (RS) 2.0
- **Q2 2002** – ebXML Collaborative Partner Profile Agreement (CPPA) 2.0
- **Q2 2002** – ebXML Registry Information Model (RIM) 2.0
- **Q2 2002** – Security Assertion Markup Language (SAML) 1.0
- **Q3 2002** – ebXML Message Service Specification 2.0
- Q1 2003 – UDDI 2.0
- **Q1 2003** – Extensible Access Control Markup Language (XACML) 1.0
- **Q2 2003** – Service Provisioning Markup Language (SPML) 1.0
- **Q2 2003** – XML Common Biometric Format (XCBF) 1.1
- **Q2 2003** – Web Services for Remote Portlets (WSRP) 1.0
- **Q3 2003** – Security Assertion Markup Language (SAML) 1.1
- **Q1 2004** – Common Alerting Protocol 1.0
- **Q1 2004** – Application Vulnerability Description Language (AVDL) 1.0
- **Q1 2004** – Web Services Security (WS-Security) 1.0, which included the following specifications:
    - SOAP Message Security 1.0 (WS-Security 2004)
    - Username Token Profile 1.0
    - X.509 Token Profile 1.0
- **Q4 2004** – WS-Security SAML Token Profile 1.0
- **Q4 2004** – WS-Security REL Token Profile 1.0
- **Q4 2004** – WS-Reliability (WS-R) 1.1
- **Q4 2004** – Universal Business Language (UBL) 1.0
- **Q1 2005** – Security Assertion Markup Language (SAML) 2.0

- **Q1 2005** – UDDI v3.02
- Q1 2005 Web Services Distributed Management (WSDM)
- **Q1 2005** – XACML v2.0
- **Q1 2005** – Revised IPR policy approved by OASIS Board of Directors
- **Q2 2005** – ebXML Registry Information Model (RIM) v3.0 and ebXML Registry Services and Protocols (RS) v3.0
- **Q2 2005** – Darwin Information Typing Architecture (DITA)
- **Q4 2005** – Web Services Security v1.1. (public review completed at the time of writing)

A.2.5 Influence and Future Directions

OASIS has a tremendous influence on e-business standards, especially in the area of XML-based Web services and related standards. These include key suites of standards for business interchange, such as ebXML, UBL and others, as well as core Web services standards, such as WS-Security, SAML, and UDDI.

OASIS drives the standardization of horizontal standards that can be used across industry groups, such as security standards for Web services. The organization also drives the development of vertical industry standards, such as industry-specific XML vocabularies (for example, the work carried out in the legalXML community). Although OASIS is well known for standards development, the organization is also committed to promoting the adoption of e-business standards. One example is the PKI member section, which is addressing issues related to the adoption of PKI technology. Other efforts include interoperability events, the OASIS Symposium, and information channels used to share information.

OASIS is driven by its membership. Therefore, the future directions of OASIS will depend on the needs of the membership. As technologies become mature, and as existing standards are deployed, OASIS is positioned to support the continuous development of new standards and to address end-user adoption issues.

A.3 WS-I

The Web Services Interoperability Organization (WS-I) is an organization chartered to "promote Web services interoperability across platforms, applications and programming languages," and to work as a "standards integrator to help Web services advance in a structured, coherent manner" (WS-I overview presentation "Interoperability: Ensuring the Success of Web Services").

A.3.1 Value Proposition and Membership Base

The goal of WS-I is to promote the accelerated deployment and adoption of Web services technology by reducing the costs and risks of end-user companies, and by enabling faster development for vendors. This is made possible by creating profiles that enable the interoperability of related specifications, by documenting best practices, and by providing test tools that enable vendors to increase productivity.

The collaboration and participation of a large number of members makes adoption more likely, which in turn reduces adoption risks. A vendor carrying the WS-I logo expresses conformance to WS-I profiles. WS-I focuses on generic Web services protocols, independent of the action indicated by a message, apart from the secure, reliable, and efficient delivery of the message. Interoperability means that the message can be sent and received by platforms using a variety of operating systems and programming languages. WS-I has approximately 130 member organizations, of which about 70% are vendors and 30% are end-user organizations. The organization has a strong non-US membership, including the highly influential Japan Special Interest Group.

A.3.2 Organization and Operating Model

The WS-I Board of Directors is "responsible for administration and operations and for ensuring that the organization adheres to its charter of delivering practical, unbiased guidance and resources and tools to promote interoperable Web services" (WS-I, "How we work").

WS-I was founded by the following nine companies: Accenture, BEA Systems, Fujitsu, Hewlett-Packard, IBM, Intel, Microsoft, Oracle, and SAP AG. The WS-I Board of Directors includes eleven members. All of the founding members except Accenture have a permanent position on the board (Accenture resigned its earlier position); two positions are elected. An election is held every year for one board position for a two-year term. For more information on the WS-I Board, visit <http://www.ws-i.org/about/leadership.aspx>.

WS-I maintains the following working groups, listed with their respective goals:

- **Basic Profile** – The Basic Profile Working Group profiles core Web service standards. Its work has included the development of the Basic Profile 1.1, which outlines the use of SOAP 1.1, WSDL 1.1, and UDDI 2, and the Attachments Profile 1.0 for the W3C SOAP Messages with Attachments (SwA) Note.
- **Basic Security Profile** – The Basic Security Working Group develops an interoperability profile for Web Services Security, focusing on SOAP Message Security (WS-Security), but also considering transport security.
- **Requirements Gathering** – The Requirements Gathering Working Group identifies interoperability issues related to Web services that are candidates for WS-I profiling work.
- **Sample Applications** – The Sample Applications Working Group creates sample applications illustrating Web services scenarios that benefit from WS-I activities.
- **Testing Tools** – The Testing Tools Working Group develops materials that can be used to test the conformance of Web service implementations to WS-I profiles.
- **XML Schema Work Plan** – The XML Schema Work Plan Working Group investigates and proposes work plans for a WS-I profile related to XML Schema interoperability.

In addition to the technical working groups, WS-I also has several special interest groups (SIGs) related to non-technical activities, such as marketing. The WS-I SIGs include the following:

- **Liaison Committee** – The Liaison Committee manages relationships with other organizations involved in standards-related work.
- **Marketing and Communications Committee (MCC)** – The MCC is responsible for all outside communications, including press and analyst relations, speaking opportunities, events, and conferences.
- **Japan Special Interest Group** – The Japan SIG consists of active Japanese member companies. It supports the organization's work in Japan, and its activities include deliverables translation, events, and media activities.

A.3.3 Deliverables

WS-I deliverables include profiles, test tools, and sample applications. Profiles outline how groups of specifications can be used to create interoperable solutions. The profiles refer to specific versions of specifications, clarify best practices and usage, and constrain the specifications where possible to enable interoperability. WS-I has produced the Basic Profile 1.1 outlining the use of SOAP 1.1, HTTP 1.0/1.1, WSDL 1.1, and UDDI 2.0, the Attachments Profile 1.0 outlining the use of SOAP Messages with Attachments, and is working on a Basic Security Profile that outlines the use of the OASIS Web Services Security standards to secure SOAP messages in an interoperable manner.

WS-I sample applications are multi-vendor demonstrations of interoperability, which use WS-I profiles of Web service standards. A realistic supply chain scenario has been implemented using components from various WS-I participants, allowing interoperability to be demonstrated through the smooth cooperation of various combinations of components. The current sample application demonstrates the Basic Profile, and work is underway to incorporate the Basic Security Profile as well.

WS-I test tools are software components that allow an organization to test an interface for conformance to WS-I profiles. The test tools include a monitor to capture protocol traces, and an analyzer to examine these traces for conformance to WS-I profiles. This makes it possible for the implementers of Web services to accelerate their deployment through simplified testing of WS-I profile conformance. Work is in progress to create test tools capable of testing the Basic Security Profile as well.

In addition, the WS-I community builds consensus and market awareness of Web services to accelerate their adoption.

A.3.4 Achievements

- **Q1 2002** – WS-I founded by nine founding members
- **Q1 2003** – Two additional (elected) board seats – Sun and webMethods elected by membership
- **Q3 2003** – Basic Profile 1.0 approved by WS-I membership. Profiled specifications include SOAP 1.1, WSDL 1.1, UDDI 2.0, XML 1.0, and XML Schema.
- **Q4 2003** – WS-I Supply Chain Management Sample Application 1.0 and related Supply Chain Management Use Cases, Sample Architecture Usage Scenarios, and Supply Chain Management Sample Architecture published.

The sample application demonstrates interoperable components involving the following ten companies: BEA Systems, Bowstreet, Corillian, IBM, Microsoft, Novell, Oracle, Quovadx, SAP, and Sun Microsystems.

- **Q1 2004** – Testing Tools 1.0 available. This includes a Web Service Communication Monitor and the Web Service Profile Analyzer. The monitor captures messages for analysis and the analyzer examines messages for conformance to the Basic Profile (as well as WSDL documents, XML Schema files, and UDDI entries).

- **Q3 2004** – Basic Profile 1.1, Attachment Profile 1.0, and Simple SOAP Binding 1.0 approved by WS-I membership.

  The Simple SOAP Binding material was originally in the Basic Profile 1.0, but it was separated to enable the Basic Profile 1.1, which can be composed with either the Attachment Profile or the Simple SOAP Binding. Conformance to the Basic Profile 1.1 and Simple SOAP Binding 1.0 is equivalent to conformance to the Basic Profile 1.0 and Basic Profile 1.0 Errata. The Attachment Profile is a profile of the W3C SOAP Messages with Attachments (SwA) packaging structure.

- **Q1 2005** – Basic Profile 1.0 , Basic Profile 1.1 and Simple SOAP Binding 1.0 Test Assertions Documents (TAD) Version 1.1 approved by WS-I membership.

- **Q2 2005** – Security Challenges, Threats and Countermeasures Version 1.0 Final Material approved by WS-I membership.

The following are expected 1Q 2006:

- Basic Security Profile v1.0 for OASIS Web Services Security v1.0 and Basic Security Profile v1.1 for OASIS WSS v1.1 approved by WS-I membership. These profiles will include the OASIS SOAP Message Security, UsernameToken, X.509, SAML, REL, and Kerberos token profiles, the ,SwA profile and SSL/TLS.

A.3.5 Influence and Future Directions

WS-I has significant influence on the implementations of Web services since end users expect – and vendors value – compliance to WS-I profiles.

WS-I compliance is becoming a means to accelerate interoperability and to quickly learn how to build an interoperable Web service implementation. WS-I has been building on the foundation of the Basic Profile by addressing the need for secure Web services. The Basic Security Profile and related test tools, as well as the sample application, help drive secure Web services interoperability. By promoting and enabling interoperability, WS-I will have long-term influence on Web services.

A.4 Liberty Alliance

The Liberty Alliance is an open organization actively involved in all aspects of identity management. Unlike most standards organizations, Liberty Alliance has since its founding in September 2001 dealt with technology, privacy, and business aspects of identity management as a unified whole. As a result, the work of the organization has emphasized business, regulatory, public policy, and legal considerations in addition technology issues.

From the start, the Liberty Alliance has strived to provide means to enable distributed management of identities, rather than focusing on a single point of control and management of identity-related information. Technical specifications are therefore needed to address the demand for *identity federation* to solve the problem of how different entities can share identity information across networks and organizational boundaries. From the very beginning, companies from within the mobile industry have been very active within the Liberty Alliance. As a result, Liberty Alliance specifications have been applicable to the mobile as well as the fixed Internet. User Identity information, attributes, and preferences constitute the core components of any online transaction.

Due to the potential privacy problems resulting from misused personal information, it is essential for any architecture dealing with personal information to meet the privacy and security requirements of users and organizations. Furthermore, it is necessary to maintain a dialogue with authorities, and make sure that the deliverables adhere to relevant data protection laws and regulations, which is another core activity within the Liberty Alliance.

A.4.1 Value Proposition and Membership Base

Businesses and consumers want to enjoy the benefits of being connected at any time and from any place, without compromising security or control of personal information. The Liberty Alliance provides the technology, knowledge, and certifications for building identity into the foundation of mobile and Web-based communications and transactions (for more information, visit <http://www.projectliberty.org/>). The Liberty Alliance is unique in being the only open organization that spans over technical, business, and policy aspects of identity management.

Another interesting aspect of the alliance is its heterogeneous membership base, which includes government organizations, IT vendors, system integrators, and end-user companies. Many of the members come from different vertical industries, such as the financial, media, gaming, ISP, manufacturing, travel, and fixed and mobile Internet industries. This strengthens the applicability of all Liberty Alliance solutions.

A.4.2 Organization and Operating Model

The Management Board is responsible for the overall governance of the Liberty Alliance, including its scope of activities, overall roadmaps, budgeting, and mission and goal setting. Board member companies for 2005 include America Online (AOL), Ericsson, Fidelity Investments, France Telecom, General Motors, Hewlett-Packard, IBM, Intel, Nokia, Novell, Oracle, RSA Security, Sun Microsystems, Verisign, and Vodafone. For more information on the Liberty Alliance Management Board, visit <https://www.projectliberty.org/about/officer_bios.php>.

The Executive Director of the Liberty Alliance is responsible for the overall co-ordination of activities, including member recruitment and member retention. In addition, he is a frequently quoted spokesperson. The Director of Operations reports to the Executive Director and is responsible for coordinating the organization's internal and external staff. The administrative staff is responsible for, for example, developer resources, technical editing, meeting facilitation, and the Liberty Alliance Web site. The Liberty Alliance also has staff to take care of recruitment and member retention.

The Business and Marketing Expert Group (BMEG), is the entity responsible for the marketing and PR activities of the Liberty Alliance. In addition to public relations, BMEG also handles other tasks via the following sub-teams:

- **The Market Requirements Documents (MRD)** sub-team collects requirements for specification work. Currently, some of the key areas include Strong Authentication and Intelligent Clients. The Market Requirements process drives all technical work in the alliance.
- **The Business Guidelines** sub-team is chartered to help organizations facilitate the business ecosystem that is needed in the distributed online environment. Deliverables include a generic guideline as well

as specific vertical documents for the mobile and 401K industries. BMEG has both its own staff and external contractors to manage events and press and analyst relations mainly across Europe, North America, and Asia.

The Technology Expert Group (TEG) is responsible for all technical work within the alliance. All work undertaken by TEG is based on MRDs delivered by BMEG. The main activities of TEG are:

- Further development and maintenance of the Identity Web Services Framework (ID-WSF), including incorporating support for standards provided by other organizations such as OASIS and the W3C.
- Technical specification development based on MRDs delivered by BMEG.
- Driving standards convergence by determining which emerging open standards are applicable to the Liberty Alliance architecture, and using them within the Liberty Alliance architecture.
- Interoperability testing to verify the validity of the specifications. One key reason behind the short time from a finished specification to the first commercial implementation is the ability to produce well-specified and interoperability-tested sets of specifications that deliver a coherent architecture. For example, AOL launched its Liberty-enabled Web service implementation within two months after the final ID-WSF1.0 specifications were announced.

The Services Group (SG) is responsible for defining Service Interfaces that use the Identity Web Services Framework (ID-WSF) architecture to further ease the deployment of specific applications. Current Identity Service Interface Specifications (ID-SIS) include:

- **Liberty ID-SIS Contact Book Service Specification** provides identity-based specifications, which help Liberty-enabled products to realize Contact Book services by using existing Contact Book protocols. It provides a method for invoking an existing Contact Book for the purpose of accessing and managing contacts. It adds privacy, security, and interoperability qualities to existing Contact Book services and protocols in Liberty-enabled environments.
- **Liberty ID-SIS Geolocation Service Specification** enables interoperable Web services information exchange to access location data, such as coordinates and/or civil address data, in a secure, privacy-protected manner.

- **Liberty ID-SIS Presence Service Specification** provides an identity-based specification designed to help Liberty-enabled products realize presence services using existing protocols. The specification outlines the privacy-friendly, secure information exchange between a presence service provider and other service providers within the same circle of trust.
- Additional service interface specifications are expected to emerge over time.

The Public Policy Expert Group (PPEG) addresses privacy and policy aspects of identity management, and is a unique entity when compared to other existing open standards organizations.

Identity management solutions handle private information, which in many cases is personally identifiable information. Thus, any attempt to provide technical specifications in this area requires a strong understanding of the associated privacy and policy requirements, and laws observed in different geographical areas. PPEG is a unique group of experienced privacy lawyers and policy experts working to ensure that the Liberty specifications offer the right set of technical features for all deployment environments. To do this, this group frequently relies on the comments and advice of various external organizations, advocacy groups, and industry experts. The scope of PPEG activities includes:

- Review of all technical work within the Liberty Alliance to ensure that they contain the right set of privacy features and to suggest additions or enhancements when needed.
- Privacy and Policy Best Practices, providing guidelines on how to implement Liberty-enabled solutions while still adhering to various privacy and data security regulations (see <http://www.projectliberty.org/specs/final_privacy_security_best_practices.pdf>)
- Managing the Liberty Alliance eGovernment initiative, supported by, for example, the government organization members of the Liberty Alliance.
- Actively participating in eGovernment initiatives on three continents: the Americas, Asia, and Europe.
- Dialogue and collaboration with public and other policy organizations.
- Pro-actively engaging with privacy and policy organizations or relevant government ministries to inform them on Liberty's deliverables, and translating their feedback to specification development.

Special interest groups (SIGs) have also been established to focus on specific topics of joint interest to a group of members, regardless of membership type. The most significant SIGs include:

- Japan SIG (previously a subteam of BMEG, but now allows for full member participation).
- Identity Theft Prevention Working Group. As a part of the initiative, collaboration with other organizations is also under way.
- An eHealth Care SIG.

In addition, the Liberty Alliance has a conformance program. Successful participation in the conformance program entitles vendors to use the Liberty Alliance Interoperable™ logo for tested products.

A.4.3 Deliverables

The Liberty Alliance has produced a technical architecture for identity management that meets business and regulatory requirements. This includes a set of technical specifications outlining an Identity Federation Framework (ID-FF) and an Identity Web Services Framework (ID-WSF) enabling identity-based Web services to be provided, discovered, and invoked while protecting privacy, as well as a set of Identity Service Interface Specifications (ID-SIS).

The specifications have been driven by business and technical requirements determined by the Liberty Alliance membership, which includes many end-user organizations.

In addition to a technical architecture and specifications, the Liberty Alliance has produced a series of business guidelines to enable organizations to understand the regulatory, legal, and business risks related to identity management. The business guidelines also describe how to address these risks through a combination of contractual, process, policy, compliance, and technical approaches. The organization has also produced documents related to privacy and to meeting privacy regulations with a Liberty Alliance deployment.

One example is the Mobile Business Guidelines document introduced in Q3 2003 (see <https://www.projectliberty.org/resources/mobile/Liberty_Business_Guidelines_Mobile_Deployments_Whitepaper_040505.pdf>), which provides guidance on managing the mutual confidence, risk, liability, and compliance aspects of a Liberty-enabled mobile-centric ecosystem. Companies active in the development of this document included France Telecom/Orange, Gemplus, Nokia, Sun Microsystems, and Vodafone.

A.4.4 Achievements

During its relatively short history, the Liberty Alliance has made several achievements:

- **Q3-2001** – Liberty Alliance founded
- **Q1-2002** – Identity Federation Framework (ID-FF) specification work commences
- **Q3-2002** – ID-FF 1.0 specifications available
- **Q2-2003** – First public interoperability demonstration, with twenty participating companies in four use cases
- **Q2-2003** – ID-FF 1.1 specifications contributed to OASIS to drive convergence and reinforce the Secure Assertion Markup Language (SAML) specifications
- **Q2-2003** – Draft Identity Web Services Framework (ID-WSF 1.0) specifications available
- **Q2-2003** – First business guidelines document published
- **Q4-2003** – Service and Conformance Expert Groups formed, Conformance program launched
- **Q4-2003** – First conformance testing event, nine products awarded the right to use the Liberty Alliance Interoperable logotype.
- **Q4-2003** – ID-FF 1.2 contributed to OASIS, key input to the SAML 2.0 specification work
- **Q4-2003** – ID-WSF 1.0 final specifications
- **Q1-2004** – First ID-WSF 1.0 implementation announced by AOL
- **Q1-2004** – Mobile business guidelines v1.0 published
- **Q1-2004** – Liberty addresses Identity Theft in a study
- **Q1/2-2004** – ID-WSF interoperability demonstrated in several events, covering both mobile and fixed/wireless broadband use cases (Radio@AOL, AOL, Nokia, and Trustgenix. In Q3-2004, Sun Microsystems also provided a back-end implementation for the service.)
- **Q2-2004** – Nine new products accepted as Liberty Alliance Interoperable

- **Q2-2004** – Collaboration with six industry bodies announced, including the Network Applications Consortium (NAC), Open Mobile Alliance (OMA), Open Security Exchange (OSE), PayCircle, SIMalliance, and the WLAN Smart Card Consortium.
- **Q3-2004** – Ten companies participate in a public interoperability demonstration, with several use cases, featuring both ID-FF and ID-WSF implementations.
- **Q4-2004** – Twelve products accepted as Liberty Alliance Interoperable. The first ID-WSF conformance test event takes place, in which Nokia, Novell, NTT, Sun Microsystems, and Trustgenix earn the right to use the Liberty Alliance Interoperable logo for their respective ID-WSF implementations.
- **Q1-2005** – Draft availability of ID-WSF 2.0, featuring, for example, support for the SAML 2.0 specifications. Release of updated mobile business guidelines document.
- **Q2-2005** – Three ID-SIS specifications released; Contact Book, Presence and Geolocation. Publication of the document *Circles of Trust: The Implications of EU Data Protection and Privacy Law for Establishing a Legal Framework for Identity Federation*. Identity Theft Prevention Working Group is announced.
- **Q3-2005** – First conformance event featuring products passing SAML 2.0 interoperability testing.
- **Q4-2005** – Draft Release 2 of ID-WSF 2.0 available

The Liberty Alliance estimates that there will be some 400 million Liberty-enabled clients and identities by the end of 2005. The following are some examples of the publicly announced deployments to date:

- **American Express:** Internal implementation to tie distributed identity domains together within the company.
- **America Online:** Commercial implementations of Liberty Alliance specifications for the AOL Production server, featuring authentication/discovery services for AOL's Radio and Photo Services. D-Link Digital Media Adaptor DSM-320 is commercially available for Broadband customers.
- **France Telecom/Orange:** Identity Federation/Single Sign-On solution to be implemented for Orange's approximately fifty million users.
- **France Telecom/Wanadoo:** Liberty ID-FF specifications used in a federated micro-payment solution available to some ten million customers.

- **Fidelity Investments:** Liberty ID-FF specifications implemented in 401K offerings, with currently more than twenty million users in thousands of organizations.
- **General Motors:** Federated identity used to simplify outsourcing and improve the efficiency of internal processes. Different portals can be accessed by means of Single Sign-On.

A.4.5 Influence and Future Directions

The Liberty Alliance is the only identity management organization focused solely on this subject, and it is the only open organization to address all aspects of the problem. The organization's work is divided into the following areas:

- **Technical specifications** – providing technical solutions to meet business requirements
- **Interoperability testing** – providing means to validate specifications
- **Conformance testing** – providing means to validate product adherence to specifications
- **Business guidelines and privacy/policy best practices** – emphasizing the business and legal aspects of identity management

As the Liberty Alliance deliverables are being adopted broadly across industries, the organization is proactively seeking collaboration with other organizations. At the time of writing, the Liberty Alliance collaborates with the following organizations:

- **Network Applications Consortium (NAC)** – NAC is an affiliate member of the Liberty Alliance. The organization works to improve interoperability and manageability in heterogeneous, virtual enterprise computing environments.
- **OASIS** – The Liberty Alliance has contributed its ID-FF 1.1 and 1.2 specifications to the OASIS SSTC to reduce fragmentation in the standards arena. It also hosted a SAML 2.0 interoperability test event for SAML 2.0 specifications (December 2004).
- **Open Mobile Alliance** – OMA refers to Liberty ID-FF specifications, including the Liberty-Enabled Client/Proxy (LECP), in its OMA Web Services Enabler Release 1.0 (see <http://www.openmobilealliance.org/release_program/owser_v10.html>).

- **Open Security Exchange (OSE)** – OSE delivers interoperable security specifications and best practice guidelines that enable organizations to mitigate risks, optimize their security postures, and enforce privacy policies more efficiently. The collaboration with Liberty Alliance results from the fact that Federated Identity plays an important role in bridging physical and IT security systems.
- **Radicchio** – Contributed all of its work, known as Trusted Transaction Roaming (t2r), to the Liberty Alliance in October 2003.
- **SIMalliance** – Liberty Alliance has a strong representation in the SIM card industry, and the collaboration with SIMalliance is therefore a natural cross-member initiative, with SIM card vendors seeking opportunities to introduce smart cards featuring support for Liberty specifications.
- **TV Anytime** – Liberty Alliance works together with TV Anytime to address the requirements of digital identity in the Digital Video Recorder (DVR) market, including issues of privacy, security, and interoperability in TV Anytime's Phase 2 specifications.
- **WLAN Smart Card Consortium** – The consortium focuses on secure mobility management by defining and promoting specifications for worldwide access to wireless LAN networks.

One of the most important goals of the Liberty Alliance is to drive the convergence of specifications and other deliverables to establish a worldwide consensus on how to enable privacy-friendly, secure, and interoperable identity management solutions. A good example of this is the contribution of the ID-FF 1.1 and 1.2 specifications to the Secure Services Technology Commitment (SSTC) technical committee of OASIS.

This effectively means that the next specification of the SSTC, SAML 2.0, is a major step towards obtaining industry-wide acceptance for one set of federated identity specifications. To achieve its ambitious goals, the Liberty Alliance intends to continue developing collaboration and to explore new opportunities to liaise with other open organizations.

Another important requirement frequently addressed by various Liberty Alliance members is that worldwide adoption should be based on open specifications, and that the requirements of end-user companies should guide the specification work. This is the only viable route towards equal-terms competition and full interoperability.

Finally, it is essential that the Liberty Alliance deliverables adhere to privacy and policy regulations as defined by, for example, EU and US authorities. All of these aspects influence the outlining of roadmaps for the Liberty Alliance. For technical activities, the work with the Identity Web Services Framework continues, through the enhancement of current specifications to include new requirements, and through adding features to comply with leading (existing and/or emerging) Web services specifications (e.g., WSS, SAML 2.0).

The ID-WSF 2.0 specifications are being finalized, and draft specifications are being posted in portions, with the main milestones including SAML 2.0 interoperability, Phase 3 ID-WSF requirements inclusion, and finally Web Services Security and WS-Addressing support. The availability of the final specification therefore depends on the availability of other ongoing specification work. However, by introducing draft ID-WSF 2.0 versions in logical and clearly distinguishable portions, vendors can start their implementation work in parallel with the finalization process. In order to further lower the bar for integration and interoperability, the organization will also continue to develop additional Identity Service Interface Specifications. At the moment, work is under way to develop a payment service interface.

The strong authentication work will also continue as a means of enhancing interoperability and lifecycle management to achieve increased security in the area of access management.

A.5 W3C® Document Notice and License

Copyright © 1994-2002 World Wide Web Consortium, (Massachusetts Institute of Technology, Institut National de Recherche en Informatique et en Automatique, Keio University). All Rights Reserved. http://www.w3.org/Consortium/Legal/

Public documents on the W3C site are provided by the copyright holders under the following license. The software or Document Type Definitions (DTDs) associated with W3C specifications are governed by the Software Notice. By using and/or copying this document, or the W3C document from which this statement is linked, you (the licensee) agree that you have read, understood, and will comply with the following terms and conditions:

Permission to use, copy, and distribute the contents of this document, or the W3C document from which this statement is linked, in any medium for any purpose and without fee or royalty is hereby granted, provided that you include the following on ALL copies of the document, or portions thereof, that you use:

1. A link or URL to the original W3C document.

2. The pre-existing copyright notice of the original author, or if it doesn't exist, a notice of the form: "Copyright © [$date-of-document] World Wide Web Consortium, (Massachusetts Institute of Technology, Institut National de Recherche en Informatique et en Automatique, Keio University). All Rights Reserved. http://www.w3.org/Consortium/Legal/" (Hypertext is preferred, but a textual representation is permitted.)

3. If it exists, the STATUS of the W3C document.

When space permits, inclusion of the full text of this NOTICE should be provided. We request that authorship attribution be provided in any software, documents, or other items or products that you create pursuant to the implementation of the contents of this document, or any portion thereof. No right to create modifications or derivatives of W3C documents is granted pursuant to this license. However, if additional requirements (documented in the Copyright FAQ) are satisfied, the right to create modifications or derivatives is sometimes granted by the W3C to individuals complying with those requirements.

THIS DOCUMENT IS PROVIDED "AS IS," AND COPYRIGHT HOLDERS MAKE NO REPRESENTATIONS OR WARRANTIES, EXPRESS OR IMPLIED, INCLUDING, BUT NOT LIMITED TO, WARRANTIES OF MERCHANTABILITY, FITNESS FOR A PARTICULAR PURPOSE, NON-INFRINGEMENT, OR TITLE; THAT THE CONTENTS OF THE DOCUMENT ARE SUITABLE FOR ANY PURPOSE; NOR THAT THE IMPLEMENTATION OF SUCH CONTENTS WILL NOT INFRINGE ANY THIRD PARTY PATENTS, COPYRIGHTS, TRADEMARKS OR OTHER RIGHTS.
COPYRIGHT HOLDERS WILL NOT BE LIABLE FOR ANY DIRECT, INDIRECT, SPECIAL OR CONSEQUENTIAL DAMAGES ARISING OUT OF ANY USE OF THE DOCUMENT OR THE PERFORMANCE OR IMPLEMENTATION OF THE CONTENTS THEREOF.

The name and trademarks of copyright holders may NOT be used in advertising or publicity pertaining to this document or its contents without specific, written prior permission. Title to copyright in this document will at all times remain with copyright holders.

This formulation of W3C's notice and license became active on April 05 1999 so as to account for the treatment of DTDs, schemas and bindings. See the older formulation for the policy prior to this date. Please see our Copyright FAQ for common questions about using materials from our site, including specific terms and conditions for packages like libwww, Amaya, and Jigsaw. Other questions about this notice can be directed to site-policy@w3.org.

Appendix B: Nokia Web Services Development API – Quick Reference

This appendix provides a quick reference to the API of the SOA for S60 platform. The following material contains complete UML class diagrams, provided by S60 3rd Edition.[1] The API consists of a set of libraries that include the following features:

1. **Service Connection** (`SENSERVCONN.lib`) – This library contains a class for creating a SOA for S60 platform client application, and allows such an application to create a connection to a Web service.
2. **Service Manager** (`SENSERVMGR.lib`) – The SOA for S60 platform architecture is administered by the Service Manager, and a class is provided to enable client applications to interact with the manager (for example, to register services and identity provider information).
3. **Service Description** (`SENSERVDESC.lib`) – This is a set of classes for managing the description of services. It includes the ability to create and manipulate new service descriptions.

4. **XML Processing** (`SENXML.lib`) – This is a set of basic classes for XML processing. It includes the SenBaseFragment and SenBaseElement classes, which can be used to ease the burden of XML processing, providing simple API for XML (SAX) and DOM-like interfaces to XML.
5. **Utilities** (`SENUTILS.lib`) – This is a set of classes providing utility functions such as SOAP processing, date utilities, GUID generation, etc.

The following sections present the UML class diagrams of the public interfaces provided by these libraries.

B.1 Service Connection Library

A connection to a Web service is established using the `CSenServiceConnection` class. Once the connection has been established, the application may simply submit messages to the service, either with the expectation of a synchronous response (`SubmitL()` method), or by implementing handlers for the asynchronous interfaces of `MSenServiceConsumer` and invoking the SendL() method.

`MSenServiceConsumer` is an interface that must be implemented by all SOA for S60 platform client applications. It contains methods pertaining events such as service messaging and errors.

`HandleMessage()`

This method is called upon the receipt of a message from a remote Web service.

By default, the messages that are passed into this method have already had the SOAP envelope and associated XML elements removed. Thus, the XML received is just the service-specific part of the SOAP. A client application can ask to provide it with the full SOAP-enveloped message by setting a flag on the connection.

Because some SOA for S60 platform client applications will need to handle more than one message, and will also need to do something more complex than just log the received message, the `HandleMessage` implementation may become an event handling routine for received messages.

`HandleError()`

The `HandleError()` method is called when the framework detects an error. Some of the errors will result from things such as the inability to access the mobile network. Other errors may be found prior to sending a message to a service.

`SetStatus()`

This callback method is triggered when the status of a connection changes. It can be used to perform processing based on such a change.

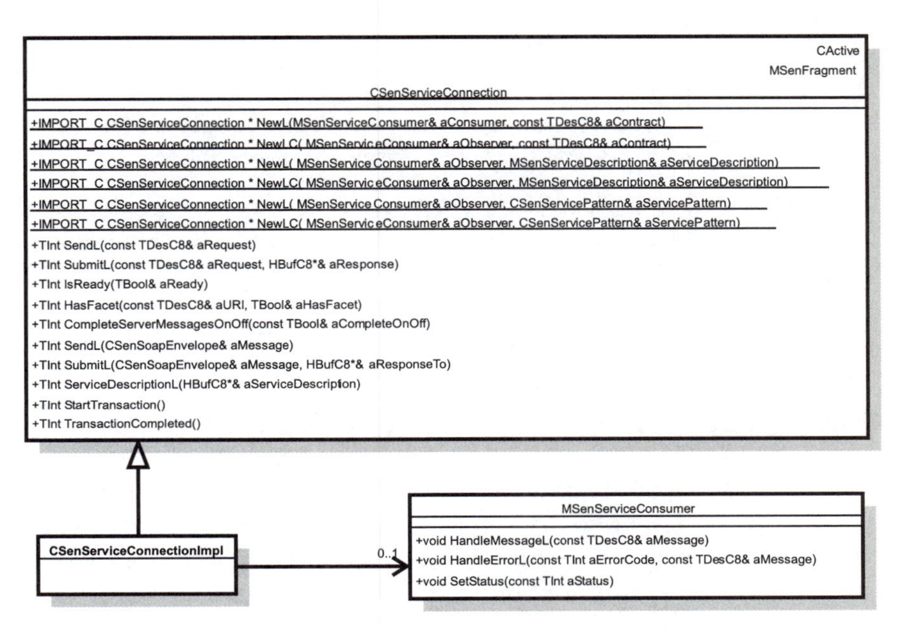

Figure B.1. Public class of the Service Connection library (SenServConn).

B.2 Service Manager Library

The SOA for S60 platform architecture is administered by the Service Manager (`CsenServiceManager`). This class, in conjunction with the `CSenServiceConnection` class, is the main entry point for SOA for S60 platform. The class is used for service and identity provider registrations and de-registrations, and also for service associations and dissociations. The actual service descriptions, however, are created using Service Description classes.

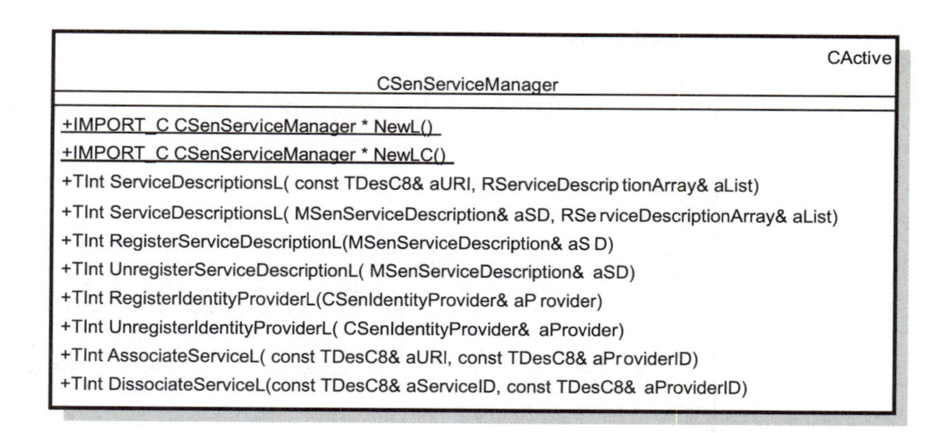

Figure B.2. Public class of the Service Manager library (SenServMgr).

B.3 Service Description Library

The Service Description library provides a set of classes for managing the description of services, including the ability to create and manipulate new service descriptions. These descriptions can then be registered into the SOA for S60 platform service database using the Service Manager library. The database stores services persistently between invocations of the SOA for S60 platform. The Service Manager reads these descriptions when it starts up, and they can then be requested by SOA for S60 platform client applications.

Service descriptions are created using SOA for S60 platform Service Description classes and then registered by using the SOA for S60 platform Service Manager.

This can be done for services that use a particular authentication service for providing an access token or a username-password combination for authentication. After registering the authentication service, or the username-password combination, it can be associated with the service. Based on this information, the SOA for S60 platform uses the selected framework to establish a connection, and if necessary, uses the authentication and discovery services to get the access credential needed for the actual service.

Once the credential has been acquired, the SOA for S60 platform caches a session with the service. If the credential is valid for a relatively long period, the caching improves performance as it reduces the number of network requests needed to access the service later. If a username-password combination is used instead of a credential, the SOA for S60 platform is nevertheless able to store the access information and to prompt the user if the provided information is not correct.

Figure B.3 and Figure B.4 illustrate the complete structure of the public interfaces of the Service Description library. A more detailed view of the main class used to manage service descriptions is shown in B.5.

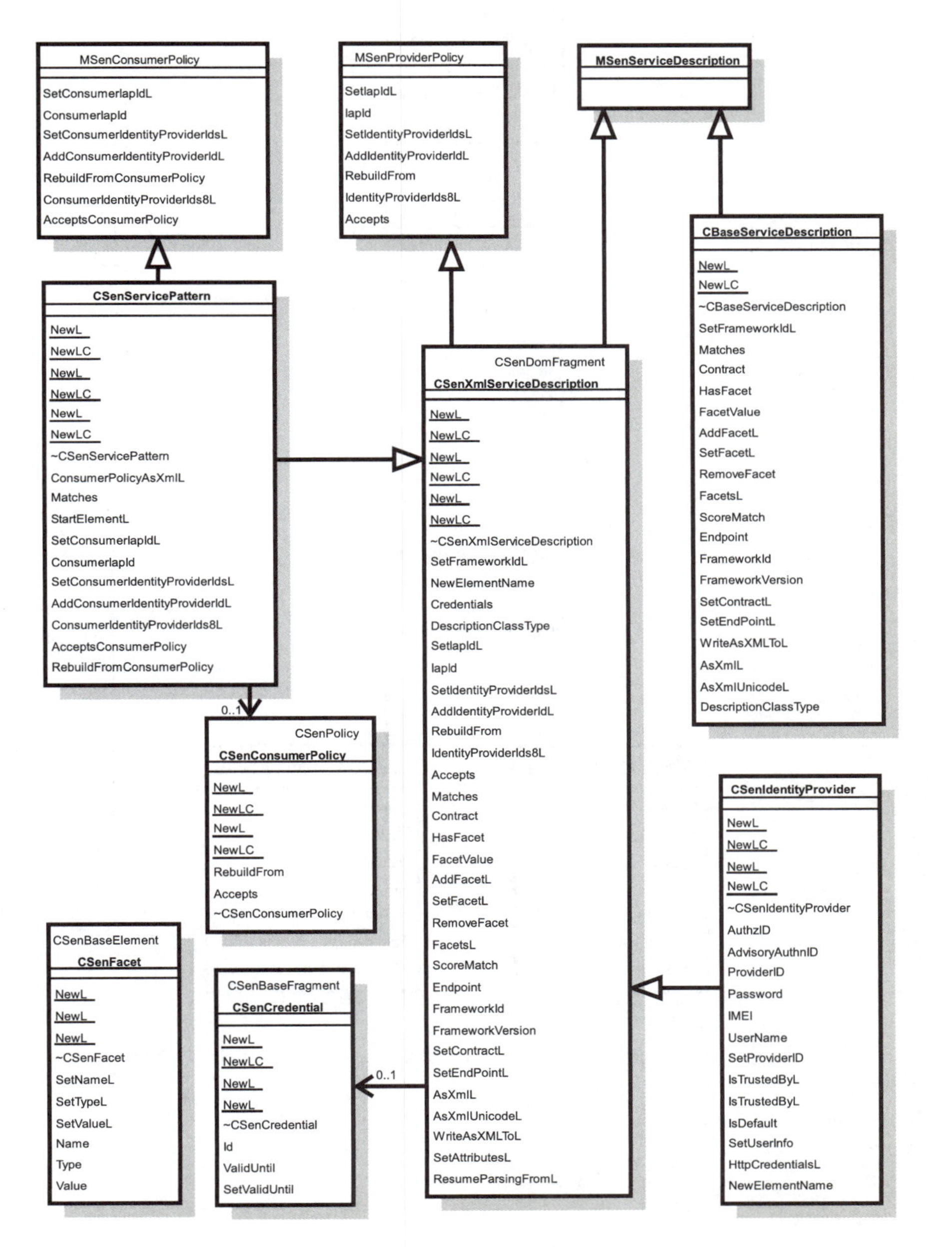

Figure B.3. Public interface of the core classes of the Service Description library (SenServDesc).

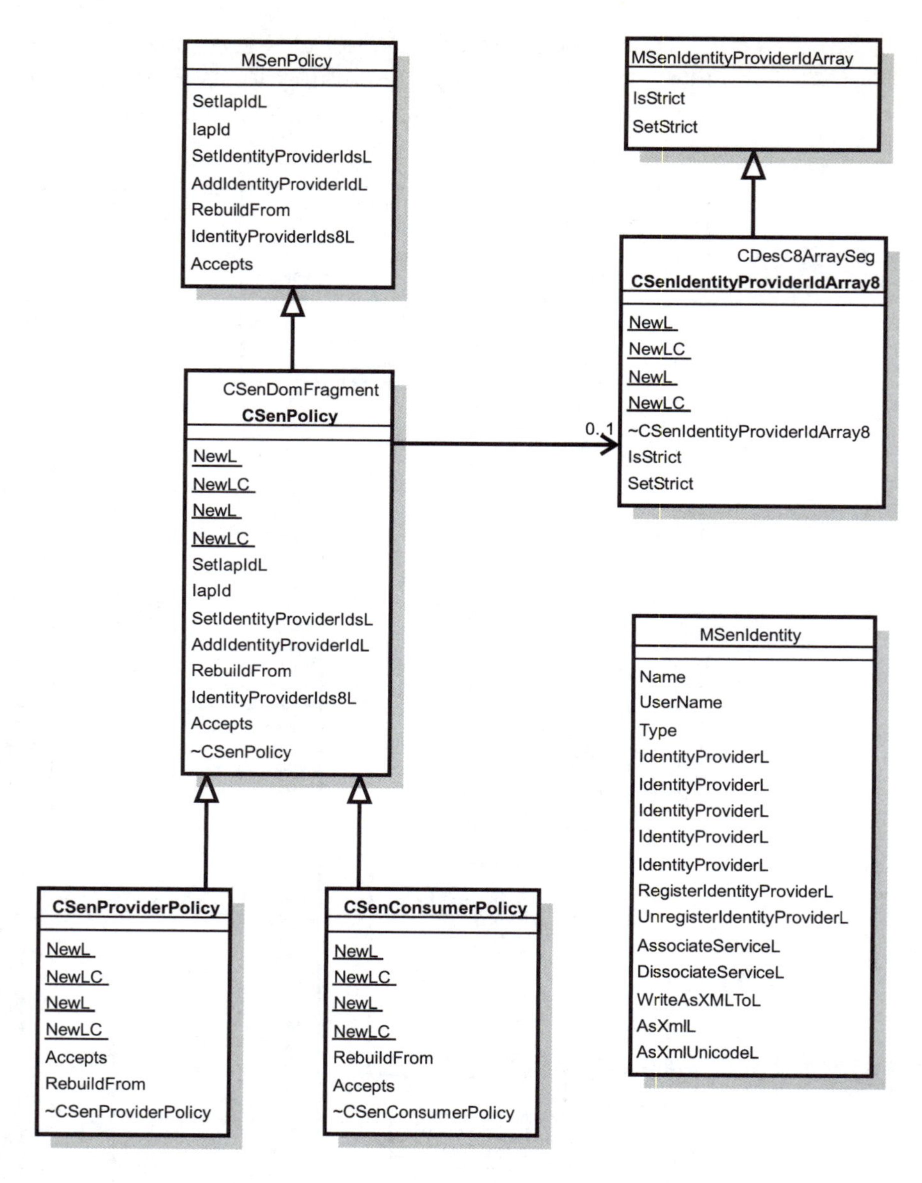

Figure B.4. Public interface of the policy-related classes of the Service Description library (SenServDesc).

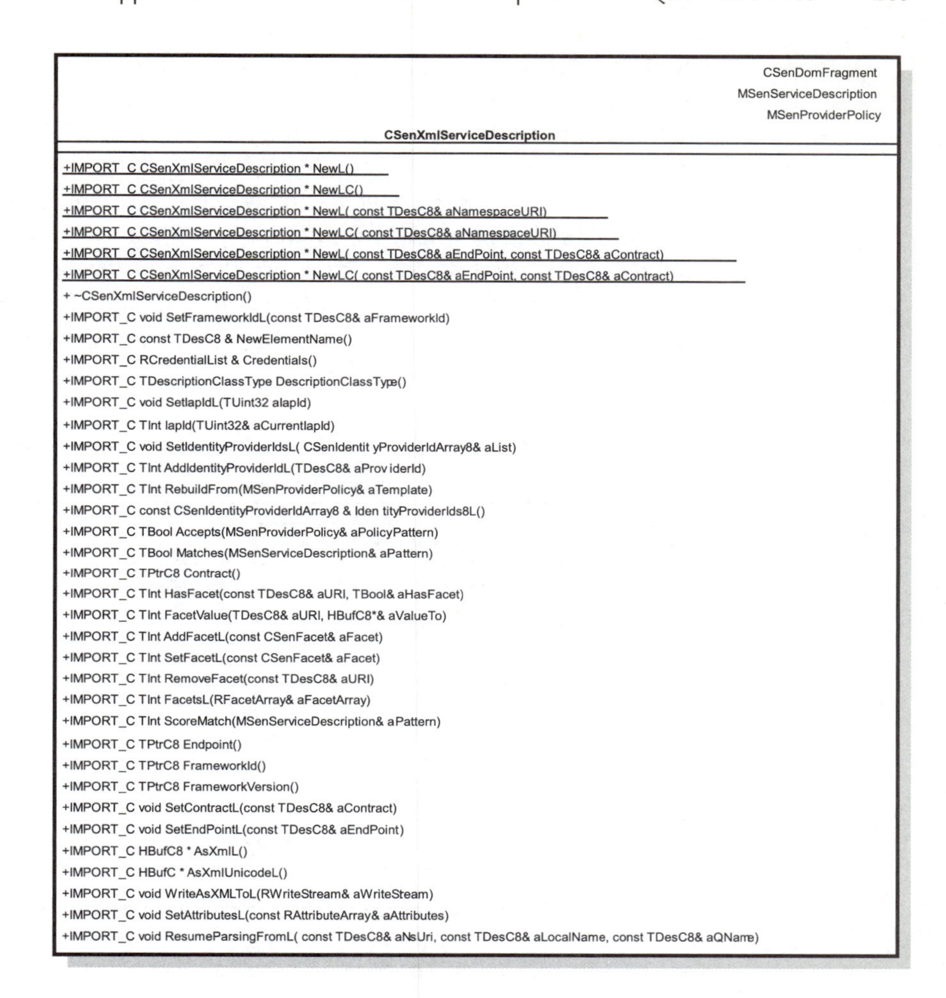

Figure B.5. Detailed view of the CSenXmlServiceDescription class of the Service Description library.

Identity is highly important in both Web services and the mobile world. Typically, Web service users (requesters) need to authenticate themselves in order to gain access to the service. The authentication process relies on either the service itself, or some other entity that the service trusts to know the identity of the service requester.

Each service may individually maintain account information for the users of the service. In such situations, the service self-authenticates the user. To support this process, the SOA for S60 platform supports authentication with WS-Security Username Token or HTTP Basic Authentication.

Services may also rely on a separate identity provider (IDP), which they trust to make assertions regarding the identity of a service-requesting user. That trust is explicitly modeled in the SOA for S60 platform, by associating IDPs with services that trust them to make such assertions. A client application that uses an IDP might register the IDP and associate itself with the IDP when the application is installed.

Services have different policies on how they authenticate a requester. The policy can vary from requiring no authentication at all to requiring authentication both for the connection from the client (peer entity), and for the actual message(s) sent by the client. Figure B-6 below provides a detailed view of `CSenIdentityProvider`, the main class used to define identity-related properties.

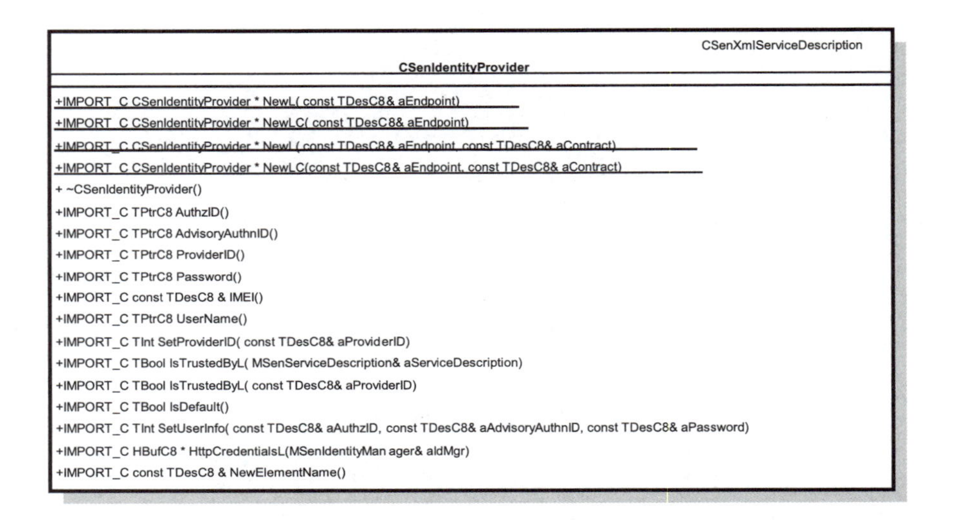

Figure B.6. Detailed view of the CSenIdentityProvider class of the Service Description library.

B.5 XML Library

The SOA for S60 platform is highly dependent on XML processing. In general, the SOA for S60 platform client applications will explicitly use XML parsing in two basic scenarios: when attempting to deal with content received as an XML message, and when producing content to be sent as an XML message.

The underlying S60 platform provides basic SAX parsing capabilities. In addition, the SOA for S60 platform offers the `SenBaseFragment` (see Figure B.8) and SenBaseElement (see Figure B.10) classes to simplify the XML parsing operations mentioned above. The SOA for S60 platform framework plug-ins and service consumer applications should use classes (such as `SenSoapMessage`) that are derived from these classes.

The basic difference between `SenBaseFragment` and `SenBaseElement` is that the fragment class relies on SAX parsing (that is, the provision of event handlers for the various SAX events) while the element class offers a simple DOM interface.

Figure B.7 below shows the complete public XML interface that is based on SAX parsing, and Figure B.9 depicts the set of classes that provide access to DOM parsing.

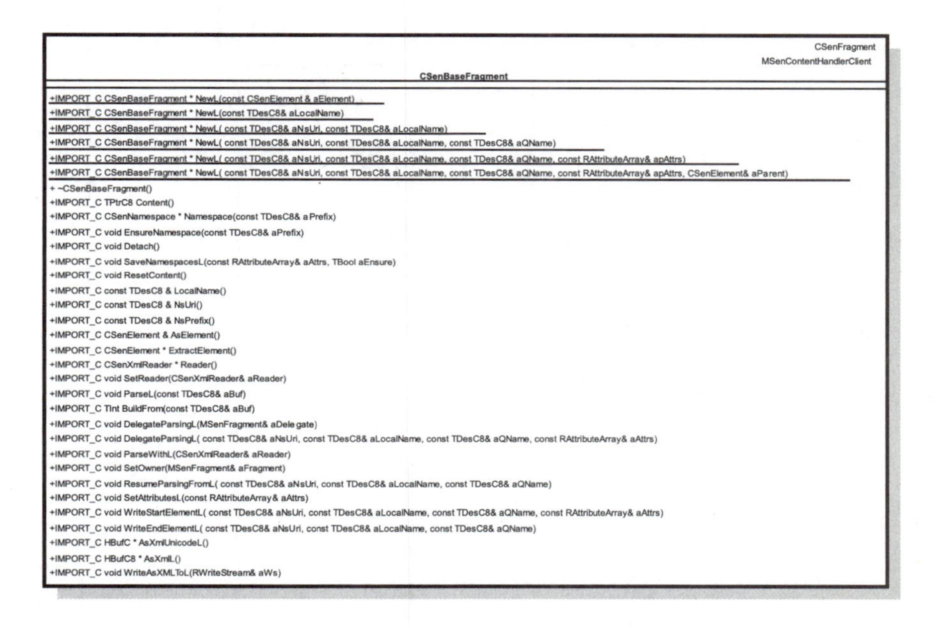

Figure B.7. Public class of SAX-related classes of the XML library (SenXml).

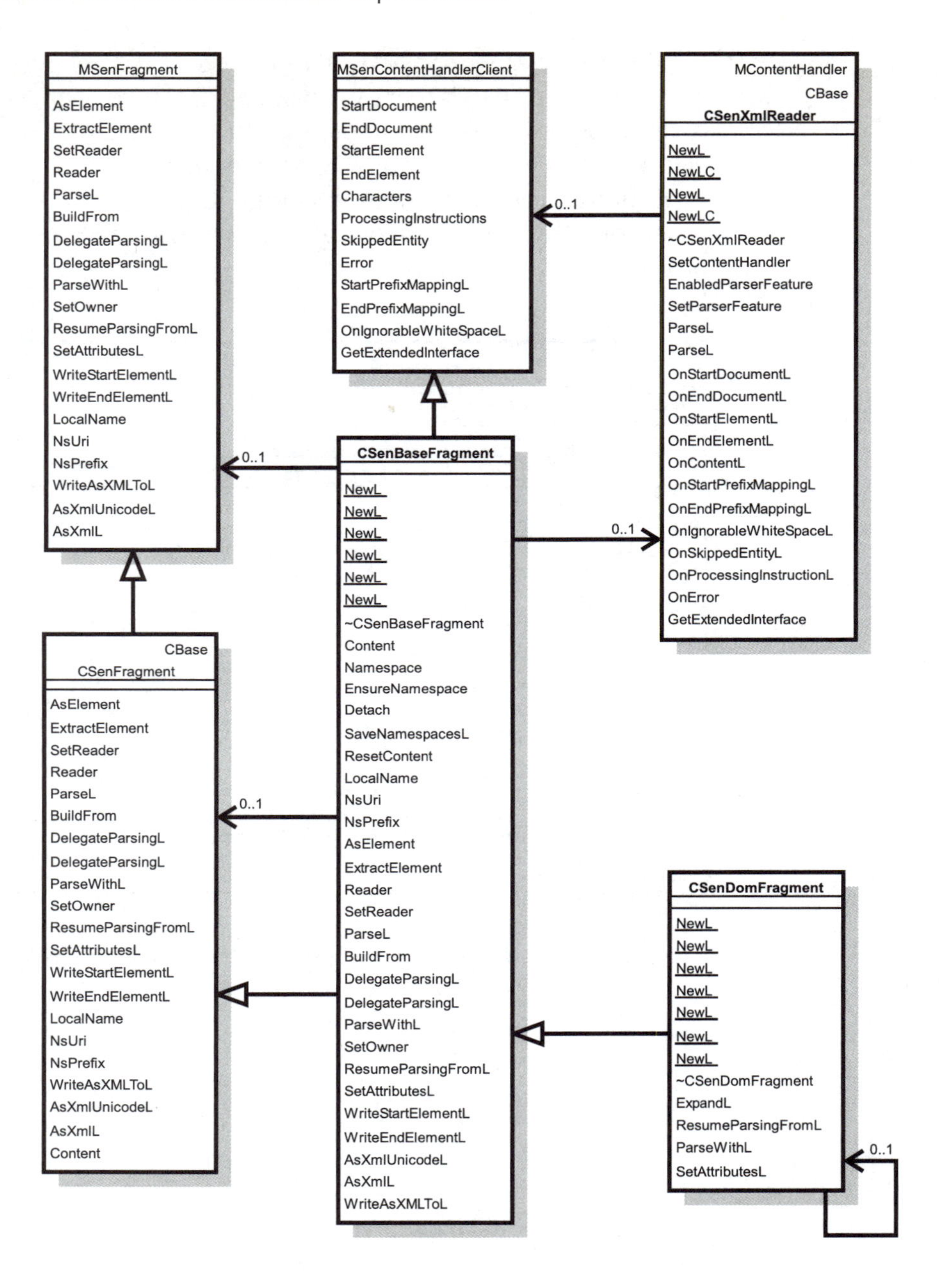

Figure B.8. The main class CSenBaseFragment used to access SAX-based parsing features.

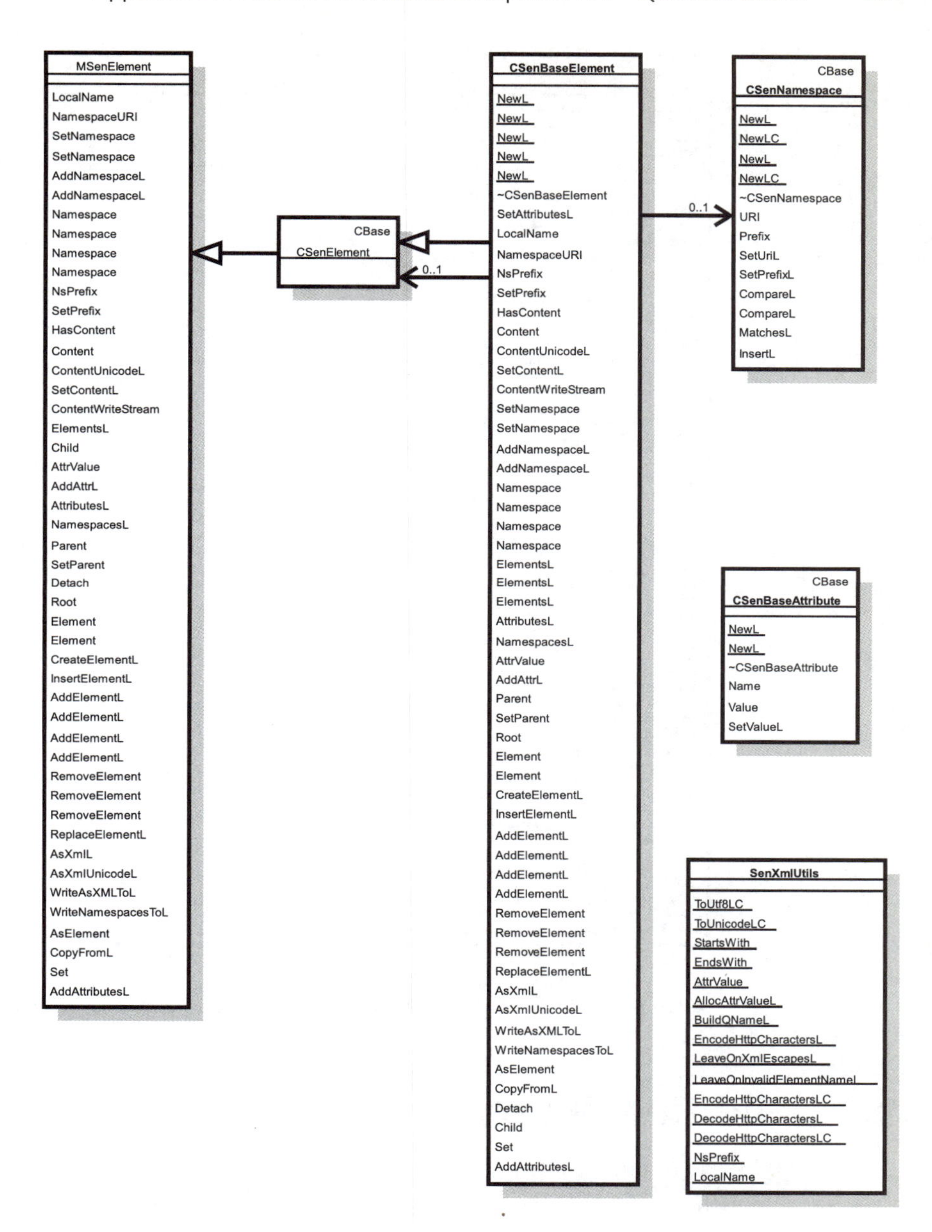

Figure B.9. Public interface of DOM-related and utility classes of the XML library (SenXml).

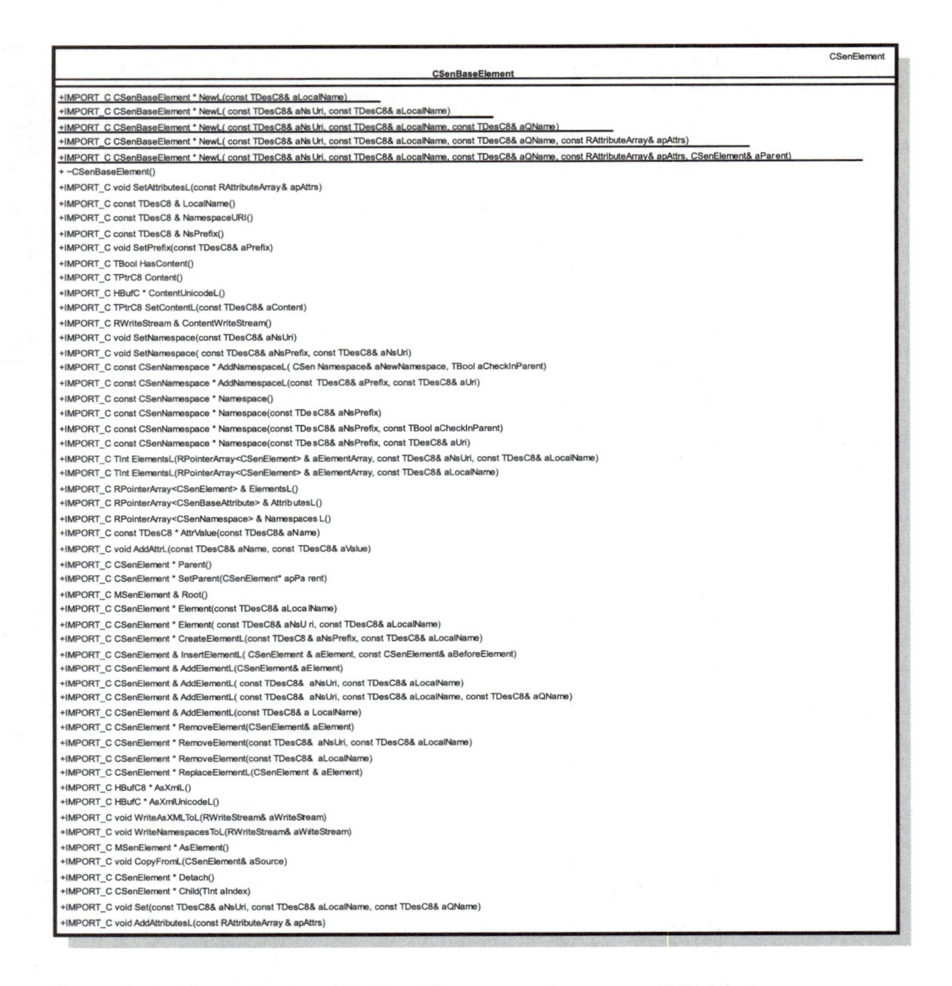

Figure B.10. The main class CSenBaseElement used to access DOM features.

B.6 Utilities Library

The Utilities library provides various helper functions, such as date formatting and the generation of unique identifiers. The main set of classes pertains to SOAP processing.

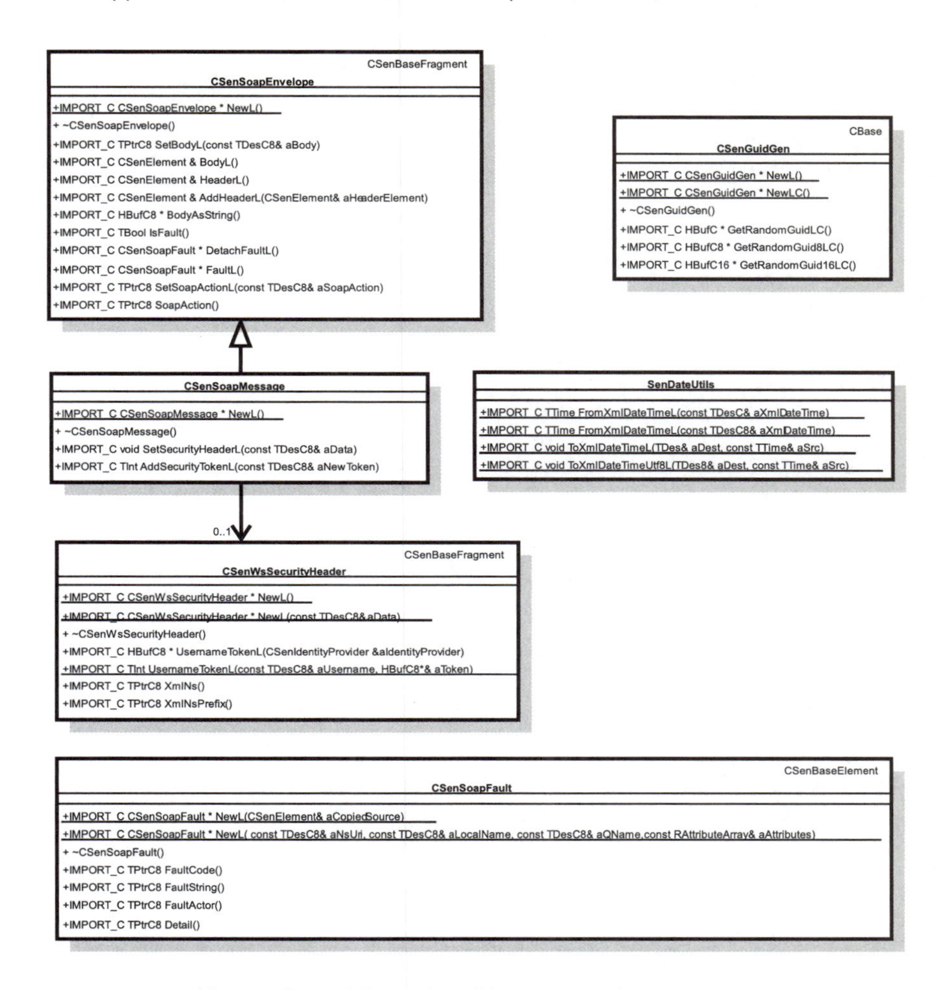

Figure B.11. Public interface of the Utilities library (SenUtils).

Appendix C: References

Chapter 2: Introduction to XML

1 Bray, Tim, et al., eds. *Extensible Markup Language (XML) 1.1 W3C Recommendation 4th February 2004 (edited in place 15 April 2004),* http://www.w3.org/TR/2004/REC-xml11-20040204/.

2 Bray, Tim, et al., eds. *Extensible Markup Language (XML) 1.1 W3C Recommendation 4th February 2004 (edited in place 15 April 2004),* http://www.w3.org/TR/2004/REC-xml11-20040204/#sec-logical-struct, sec. 3 Logical Structures.

3 Bray, Tim, et al., eds. *Extensible Markup Language (XML) 1.1 W3C Recommendation 4th February 2004 (edited in place 15 April 2004),* http://www.w3.org/TR/2004/REC-xml11-20040204/#attdecls, sec. 3.3 Attribute-List Declarations.

4 Bray, Tim, et al., eds. *Namespaces in XML 1.1 W3C Recommendation 4th February 2004,* http://www.w3.org/TR/xml-names11/.

5 Bray, Tim, et al., eds. *Namespaces in XML 1.1 W3C Recommendation 4th February 2004,* http://www.w3.org/TR/xml-names11/#uniqAttrs, sec. 6.3 Uniqueness of Attributes.

6 Bray, Tim, et al., eds. *Extensible Markup Language (XML) 1.1 W3C Recommendation 4th February 2004 (edited in place 15 April 2004),* http://www.w3.org/TR/2004/REC-xml11-20040204/.

7 Bray, Tim, et al., eds. *Extensible Markup Language (XML) 1.1 W3C Recommendation 4th February 2004 (edited in place 15 April 2004),* http://www.w3.org/TR/2004/REC-xml11-20040204/#sec-prolog-dtd, sec. 2.8 Prolog and Document Type Declaration.

8 XML Schema is defined in the following three W3C Recommendations: Fallside, David C., and Priscilla Walmsley (2nd ed.), eds. *XML Schema Part 0: Primer Second Edition W3C Recommendation 28 October 2004,* http://www.w3.org/TR/2004/REC-xmlschema-0-20041028/.

Beech, David, et al., eds. *XML Schema Part 1: Structures Second Edition W3C Recommendation 28 October 2004,* http://www.w3.org/TR/2004/REC-xmlschema-1-20041028/.

Biron, Paul V., and Ashok Malhotra, eds. *XML Schema Part 2: Datatypes Second Edition W3C Recommendation 28 October 2004,* http://www.w3.org/TR/2004/REC-xmlschema-2-20041028/.

9 Biron, Paul V., and Ashok Malhotra, eds. *XML Schema Part 2: Datatypes Second Edition W3C Recommendation 28 October 2004,* http://www.w3.org/TR/2004/REC-xmlschema-2-20041028/#built-in-datatypes, sec. 3 Built-in datatypes.

10 Beech, David, et al., eds. *XML Schema Part 1: Structures Second Edition W3C Recommendation 28 October 2004,* http://www.w3.org/TR/2004/REC-xmlschema-1-20041028/#Model_Groups, sec. 3.8 Model Groups.

11 Fallside, David C., and Priscilla Walmsley (2nd ed.), eds. *XML Schema Part 0: Primer Second Edition W3C Recommendation 28 October 2004,* http://www.w3.org/TR/2004/REC-xmlschema-0-20041028/#OccurrenceConstraints, sec. 2.2.1 Occurrence Constraints.

12 Fallside, David C., and Priscilla Walmsley (2nd ed.), eds. *XML Schema Part 0: Primer Second Edition W3C Recommendation 28 October 2004,* http://www.w3.org/TR/2004/REC-xmlschema-0-20041028/#NS, sec. 3 Advanced Concepts I: Namespaces, Schemas & Qualification.

13 Fallside, David C., and Priscilla Walmsley (2nd ed.), eds. *XML Schema Part 0: Primer Second Edition W3C Recommendation 28 October 2004,* http://www.w3.org/TR/2004/REC-xmlschema-0-20041028/#UnqualLocals, sec. 3.1 Target Namespaces & Unqualified Locals.

14 Fallside, David C., and Priscilla Walmsley (2nd ed.), eds. *XML Schema Part 0: Primer Second Edition W3C Recommendation 28 October 2004,* http://www.w3.org/TR/2004/REC-xmlschema-0-20041028/#SchemaInMultDocs, sec. 4.1 A Schema in Multiple Documents.

15 Fallside, David C., and Priscilla Walmsley (2nd ed.), eds. *XML Schema Part 0: Primer Second Edition W3C Recommendation 28 October 2004,* http://www.w3.org/TR/2004/REC-xmlschema-0-20041028/#import, sec. 5.4 Importing Types.

16 Fallside, David C., and Priscilla Walmsley (2nd ed.), eds. *XML Schema Part 0: Primer Second Edition W3C Recommendation 28 October 2004,* http://www.w3.org/TR/2004/REC-xmlschema-0-20041028/#UnqualLocals, sec. 3.1 Target Namespaces & Unqualified Locals,

Fallside, David C., and Priscilla Walmsley (2nd ed.), eds. *XML Schema Part 0: Primer Second Edition W3C Recommendation 28 October 2004,* http://www.w3.org/TR/2004/REC-xmlschema-0-20041028/#QualLocals, sec. 3.2 Qualified Locals.

Chapter 3: Introduction to Service-Oriented Architectures

Additional references are provided at the Appendix A section.

Liberty Alliance. Liberty Authentication Context Specification, Version 1.2. <http://www.projectliberty.org/specs/draft-liberty-authentication-context-1.2-errata-v1.0.pdf>.

Liberty Alliance. Liberty ID-WSF Authentication Service Specification, Version 1.0. <http://www.projectliberty.org/specs/liberty-idwsf-authn-svc-v1.0.pdf>.

Liberty Alliance. Liberty ID-WSF Discovery Service Specification, Version 1.1. <https://www.projectliberty.org/specs/liberty-idwsf-disco-svc-v1.1.pdf>.

Liberty Alliance. Liberty Reverse HTTP Binding for SOAP Specification, Version 1.0.<https://www.projectliberty.org/specs/draft-liberty-paos-1.0-errata-v1.0.pdf>.

OASIS. Web Services Security (WSS) Technical Committee. <http://www.oasis-open.org/committees/tc_home.php?wg_abbrev=wss>.

OASIS. Security Services (SAML) Technical Committee. <http://www.oasis-open.org/committees/tc_home.php?wg_abbrev=security>.

Rouault, Jason (Jan 2004). Liberty Alliance: What is Phase 2? Presentation available at <https://www.projectliberty.org/resources/presentations/Liberty%20Phase%202%20Overview.pdf>.

W3C. SOAP Messages with Attachments. W3C Note 11 December 2000. <http://www.w3.org/TR/SOAP-attachments>.

W3C. Web Services Architecture. W3C Working Group Note 11 February 2004. <http://www.w3.org/TR/ws-arch/>.

W3C. XML Encryption Syntax and Processing. W3C Recommendation 10 December 2002.<http://www.w3.org/TR/xmlenc-core/>.

W3C. XML-Signature Syntax and Processing. W3C Recommendation 12 February 2002.<http://www.w3.org/TR/xmldsig-core/>.

WS-I. Basic Profile. <http://www.ws-i.org/deliverables/workinggroup.aspx?wg=basicprofile>.

WS-I. Basic Security Profile. <http://www.ws-i.org/deliverables/workinggroup.aspx?wg=basicsecurity>.

WS-I. Supply Chain Management Sample Application Architecture. <http://www.wsi.org/SampleApplications/SupplyChainManagement/2003-12/SCMArchitecture1.01.pdf>.

1 Booth, David, et al., eds. *Web Services Architecture W3C Working Group Note 11 February 2004,* http://www.w3.org/TR/ws-arch/.

2 Christensen, Erik, et al. *Web Services Description Language (WSDL) 1.1 W3C Note 15 March 2001,* http://www.w3.org/TR/2001/NOTE-wsdl-20010315.

3 Christensen, Erik, et al. *Web Services Description Language (WSDL) 1.1 W3C Note 15 March 2001,* http://www.w3.org/TR/2001/NOTE-wsdl-20010315#_document-s, sec. 2.1 WSDL Document Structure.

4 Christensen, Erik, et al. *Web Services Description Language (WSDL) 1.1 W3C Note 15 March 2001,* http://www.w3.org/TR/2001/NOTE-wsdl-20010315#_types, sec. 2.2 Types.

5 Christensen, Erik, et al. *Web Services Description Language (WSDL) 1.1 W3C Note 15 March 2001,* http://www.w3.org/TR/2001/NOTE-wsdl-20010315#_messages, sec. 2.3 Messages.

6 Christensen, Erik, et al. *Web Services Description Language (WSDL) 1.1 W3C Note 15 March 2001,* http://www.w3.org/TR/2001/NOTE-wsdl-20010315#_porttypes, sec. 2.4 Port Types.

7 Christensen, Erik, et al. *Web Services Description Language (WSDL) 1.1 W3C Note 15 March 2001,* http://www.w3.org/TR/2001/NOTE-wsdl-20010315#_bindings, sec. 2.5 Bindings.

8 Box, Don, et al. *Simple Object Access Protocol (SOAP) 1.1 W3C Note 08 May 2000,* http://www.w3.org/TR/2000/NOTE-SOAP-20000508.

9 Box, Don, et al. *Simple Object Access Protocol (SOAP) 1.1 W3C Note 08 May 2000,* http://www.w3.org/TR/2000/NOTE-SOAP-20000508/#_toc478383532, sec. 7. Using SOAP for RPC.

10 Box, Don, et al. *Simple Object Access Protocol (SOAP) 1.1 W3C Note 08 May 2000,* http://www.w3.org/TR/2000/NOTE-SOAP-20000508/#_Toc478383528, sec. 6.1.1 The SOAPAction HTTP Header Field.

11 Christensen, Erik, et al. *Web Services Description Language (WSDL) 1.1 W3C Note 15 March 2001,* http://www.w3.org/TR/2001/NOTE-wsdl-20010315#_soap:body, sec. 3.5 soap:body.

12 Christensen, Erik, et al. *Web Services Description Language (WSDL) 1.1 W3C Note 15 March 2001,* http://www.w3.org/TR/2001/NOTE-wsdl-20010315#_ports, sec. 2.6 Ports.

13 Christensen, Erik, et al. *Web Services Description Language (WSDL) 1.1 W3C Note 15 March 2001,* http://www.w3.org/TR/2001/NOTE-wsdl-20010315#_services, sec. 2.7 Services.

14 Clement, Luc (ed.), et al., *UDDI Version 3.0.2 UDDI Spec Technical Committee Draft (Dated 20041019),* http://www.oasis-open.org/committees/uddi-spec/doc/spec/v3/uddi-v3.0.2-20041019.htm.

15 Clement, Luc (ed.), et al., *UDDI Version 3.0.2 UDDI Spec Technical Committee Draft (Dated 20041019),* http://www.oasis-open.org/committees/uddi-spec/doc/spec/v3/uddi-v3.0.2-20041019.htm#_Toc85907991, sec. 1.6.5 Taxonomic Classification of the UDDI entities.

16 *www.unspsc.org and www.naics.com*

17 Clement, Luc (ed.), et al., *UDDI Version 3.0.2 UDDI Spec Technical Committee Draft (Dated 20041019),* http://www.oasis-open.org/committees/uddi-spec/doc/spec/v3/uddi-v3.0.2-20041019.

htm#_Toc85907991, sec. 1.6.5 Taxonomic Classification of the UDDI entities.

www.gs1.org/index.php?http://www.ean-int.org/locations.html&2

18 *Introduction to UDDI: Important Features and Functional Concepts,* http://uddi.org/pubs/uddi-tech-wp.pdf, Figure 4: Conceptual Illustration of Registry Affiliation.

19 Clement, Luc (ed.), et al., *UDDI Version 3.0.2 UDDI Spec Technical Committee Draft (Dated 20041019),* http://www.oasis-open.org/committees/uddi-spec/doc/spec/v3/uddi-v3.0.2-20041019.htm#_Toc85908007, sec. 3 UDDI Registry Data Structures.

20 Clement, Luc (ed.), et al., *UDDI Version 3.0.2 UDDI Spec Technical Committee Draft (Dated 20041019),* http://www.oasis-open.org/committees/uddi-spec/doc/spec/v3/uddi-v3.0.2-20041019.htm#_Toc85908075, sec. 5 UDDI Programmers APIs.

21 Clement, Luc (ed.), et al., *UDDI Version 3.0.2 UDDI Spec Technical Committee Draft (Dated 20041019),* http://www.oasis-open.org/committees/uddi-spec/doc/spec/v3/uddi-v3.0.2-20041019.htm#_Toc85908076, sec. 5.1 Inquiry API Set.

22 Clement, Luc (ed.), et al., *UDDI Version 3.0.2 UDDI Spec Technical Committee Draft (Dated 20041019),* http://www.oasis-open.org/committees/uddi-spec/doc/spec/v3/uddi-v3.0.2-20041019.htm#_Toc85908095, sec. 5.2 Publication API Set.

23 Clement, Luc (ed.), et al., *UDDI Version 3.0.2 UDDI Spec Technical Committee Draft (Dated 20041019),* http://www.oasis-open.org/committees/uddi-spec/doc/spec/v3/uddi-v3.0.2-20041019.htm#_Toc85908128, sec. 5.5 Subscription API Set.

Chapter 4: Agreement

[SecurityChallenges]

Security Challenges, Threats and Countermeasures Version 1.0, Final Material, Date: 2005/05/07

http://www.ws-i.org/Profiles/BasicSecurity/SecurityChallenges-1.0.pdf

[WSI-BP]

Basic Profile Version 1.1, Final Material, 2004-08-24

http://www.ws-i.org/Profiles/BasicProfile-1.1-2004-08-24.html

[WSPL]

An Introduction to the Web Services Policy Language (WSPL), Anne H. Anderson

http://research.sun.com/projects/xacml/Policy2004.pdf

[WS-Policy]

Web Services Policy Framework (WS-Policy) September 2004

ftp://www6.software.ibm.com/software/developer/library/ws-policy.pdf

[WS-PolicyAttachment]

Web Services Policy Attachment (WS-PolicyAttachment) September 2004

ftp://www6.software.ibm.com/software/developer/library/ws-polat.pdf

[WS-RM-Policy]

Web Services Reliable Messaging Policy Assertion (WS-RM Policy), February 2005

http://specs.xmlsoap.org/ws/2005/02/rm/WS-RMPolicy.pdf

[WS-Security]

Web Services Security: SOAP Message Security 1.0 (WS-Security 2004) OASIS Standard 200401, March 2004

http://docs.oasis-open.org/wss/2004/01/oasis-200401-wss-soap-message-security-1.0.pdf

[WS-SecurityPolicy]

Web Services Security Policy Language (WS-SecurityPolicy) July 2005, Version 1.1

ftp://www6.software.ibm.com/software/developer/library/ws-secpol.pdf

[XACML] – (Standard)

http://www.oasis-open.org/committees/tc_home.php?wg_abbrev=xacml

1 *WS-Addressing, W3C Member Submission http://www.w3.org/2002/ws/addr/*

2 *WS-Message Delivery, W3C Member Submission http://www.w3.org/2002/ws/addr/*

3 *W3C WS-Addressing Working Group http://www.w3.org/2002/ws/addr/*

4 *WS-Addressing Core, W3C Candidate Recommendation 17 August 2005, http://www.w3.org/TR/ws-addr-core/*

5 *WS-Addressing SOAP Binding, W3C Candidate Recommendation 17 August 2005, http://www.w3.org/TR/ws-addr-soap/*

6 *WS-Addressing WSDL Binding, W3C Candidate Recommendation 17 August 2005, http://www.w3.org/TR/ws-addr-wsdl/*

7 *Internationalized Resource Identifiers (IRIs) M. Duerst, M. Suignard, January 2005. Available at http://www.ietf.org/rfc/rfc3987.txt*

8 *SOAP Version 1.2 Part 2: Adjuncts, M. Gudgin, M. Hadley, N. Mendelsohn, J-J. Moreau, H. Frystyk Nielsen, Editors. World Wide Web Consortium, 24 June 2003. http://www.w3.org/TR/soap12-part2/.*

Chapter 5: Identity and Security

Additional references are provided at the end of Appendix A.

Liberty Alliance. Liberty Authentication Context Specification, Version 1.2. <http://www.projectliberty.org/specs/draft-liberty-authentication-context-1.2-errata-v1.0.pdf>.

Liberty Alliance. Liberty ID-WSF Authentication Service Specification, Version 1.0. <http://www.projectliberty.org/specs/liberty-idwsf-authn-svc-v1.0.pdf>.

Liberty Alliance. Liberty ID-WSF Discovery Service Specification, Version 1.1. <https://www.projectliberty.org/specs/liberty-idwsf-disco-svc-v1.1.pdf>.

Liberty Alliance. Liberty Reverse HTTP Binding for SOAP Specification, Version 1.0. <https://www.projectliberty.org/specs/draft-liberty-paos-1.0-errata-v1.0.pdf>.

OASIS. Web Services Security (WSS) Technical Committee. <http://www.oasis-open.org/committees/tc_home.php?wg_abbrev=wss>.

OASIS. Security Services (SAML) Technical Committee. <http://www.oasis-open.org/committees/tc_home.php?wg_abbrev=security>.

Rouault, Jason (Jan 2004). Liberty Alliance: What is Phase 2? Presentation available at<https://www.projectliberty.org/resources/presentations/Liberty%20Phase%202%20Overview.pdf>.

W3C. SOAP Messages with Attachments. W3C Note 11 December 2000. <http://www.w3.org/TR/SOAP-attachments>.

W3C. Web Services Architecture. W3C Working Group Note 11 February 2004. <http://www.w3.org/TR/ws-arch/>.

W3C. XML Encryption Syntax and Processing. W3C Recommendation 10 December 2002. <http://www.w3.org/TR/xmlenc-core/>.

W3C. XML-Signature Syntax and Processing. W3C Recommendation 12 February 2002. <http://www.w3.org/TR/xmldsig-core/>.

WS-I. Basic Profile. <http://www.ws-i.org/deliverables/workinggroup.aspx?wg=basicprofile>.

WS-I. Basic Security Profile. <http://www.ws-i.org/deliverables/workinggroup.aspx?wg=basicsecurity>.

WS-I. Supply Chain Management Sample Application Architecture. <http://www.wsi.org/SampleApplications/SupplyChainManagement/2003-12/SCMArchitecture1.01.pdf>.

Chapter 6: Liberty Alliance Identity Technologies

1 *www.projectliberty.org*

2 Cantor, Scott (Ed.), et al., *Liberty ID-FF Protocols and Schema Specification,* http://www.projectliberty.org/specs/draft-liberty-idff-protocols-schema-1.2-errata-v3.0.pdf, sec. 3.2. Single Sign-On and Federation Protocol.

3 Ibid., sec. 3.2.1.1. Element <AuthnRequest>.

4 Cantor, Scott (Ed.), et al., *Liberty ID-FF Bindings and Profiles Specification,* http://www.projectliberty.org/specs/draft-liberty-idff-bindings-profiles-1.2-errata-v2.0.pdf.

5 Cantor, Scott (Ed.), et al., *Liberty ID-FF Protocols and Schema Specification,* http://www.projectliberty.org/specs/draft-liberty-idff-protocols-schema-1.2-errata-v3.0.pdf, sec. 3.2.2.1. Element <AuthnResponse>.

6 Ibid., sec. 3.5. Single Logout Protocol.

7 Ibid., sec. 3.5.1.1. Element <LogoutRequest>.

8 Ibid., sec. 3.5. Single Logout Protocol – Line 1385.

9 Ibid., sec. 3.5.2.1. Element <LogoutResponse>.

10 Ibid., sec. 3.4. Single Federation Termination Notification Protocol.

11 Ibid., sec. 3.4.1.1. Element <FederationTerminationNotification>.

12 Ibid., sec. 3.3. Name Registration Protocol.

13 Ibid., sec. 3.3.1.1. Element <RegisterNameIdentifierRequest>.

14 Ibid., sec. 3.3. Name Registration Protocol – Line 1183, Ibid., sec. 3.3. Name Registration Protocol - Line 1191.

15 Cantor, Scott (Ed.), et al., *Liberty ID-FF Protocols and Schema Specification,* http://www.projectliberty.org/specs/draft-liberty-idff-protocols-schema-1.2-errata-v3.0.pdf, sec. 3.3. Name Registration Protocol - Line 1187.

16 Ibid., sec. 3.3.2.1. Element <RegisterNameIdentifierResponse>.

17 Ibid., sec. 3.6. Name Identifier Mapping Protocol.

18 Ibid., sec. 3.6.1.1. Element <NameIdentifierMappingRequest>.

19 Ibid., sec. 3.6.2.1. Element <NameIdentifierMappingResponse>.

20 Cantor, Scott (Ed.), et al., *Liberty ID-FF Bindings and Profiles Specification,* http://www.projectliberty.org/specs/draft-liberty-idff-bindings-profiles-1.2-errata-v2.0.pdf., sec. 3.2. Single Sign-On and Federation Profiles.

21 Ibid., sec. 3.2.2. Liberty Artifact Profile.

22 Ibid., sec. 3.2.3. Liberty Post Profile.

23 Ibid., sec. 3.2.4. Liberty-Enabled Client and Proxy Profile.

24 Ibid., sec. 3.5. Single Logout Profiles.

25 Ibid., sec. 3.5.1.1.1. HTTP-Redirect Implementation (initiated at IDP), Ibid., sec. 3.5.2.1. HTTP-Based Profile (initiated at SP).

26 Cantor, Scott (Ed.), et al., *Liberty ID-FF Bindings and Profiles Specification,* http://www.projectliberty.org/specs/draft-liberty-idff-bindings-profiles-1.2-errata-v2.0.pdf., sec. 3.5.1.2. SOAP/HTTP-Based Profile (initiated at IDP), Ibid., sec. 3.5.2.2. SOAP/HTTP-Based Profile (initiated at SP).

27 Cantor, Scott (Ed.), et al., *Liberty ID-FF Bindings and Profiles Specification,* http://www.projectliberty.org/specs/draft-liberty-idff-bindings-profiles-1.2-errata-v2.0.pdf., sec. 3.5.1.1.2. HTTP-GET Implementation.

28 Ibid., sec. 3.4. Identity Federation Termination Notification Profiles.

29 Ibid., sec. 3.4.1.1. HTTP-Redirect-Based Profile (Initiated at IDP), Ibid., sec. 3.4.2.1. HTTP-Redirect-Based Profile (Initiated at SP), Ibid., sec. 3.4.1.2. SOAP/HTTP-Based Profile (Initiated at IDP), Ibid., sec. 3.4.2.2. SOAP/HTTP-Based Profile (Initiated at SP).

30 Cantor, Scott (Ed.), et al., *Liberty ID-FF Bindings and Profiles Specification,* http://www.projectliberty.org/specs/draft-liberty-idff-bindings-profiles-1.2-errata-v2.0.pdf., sec. 3.3. Register Name Identifier Profiles.

31 Ibid., sec. 3.3.1.1. HTTP-Redirect-Based Profile (Initiated at IDP), Ibid., sec. 3.3.2.1. HTTP-Redirect-Based Profile (Initiated at SP), Ibid., sec. 3.3.1.2. SOAP/HTTP-Based Profile (Initiated at IDP), Ibid., sec. 3.3.2.2. SOAP/HTTP-Based Profile (Initiated at SP).

32 Cantor, Scott (Ed.), et al., *Liberty ID-FF Bindings and Profiles Specification,* http://www.projectliberty.org/specs/draft-liberty-idff-bindings-profiles-1.2-errata-v2.0.pdf., sec. 3.7. NameIdentifier Mapping Profile.

33 Ibid., sec. 3.7.1. SOAP-based NameIdentifier Mapping.

34 Ibid., sec. 3.8. NameIdentifier Encryption Profile.

35 Ibid., sec. 3.8.1. XMLEncryption-based NameIdentifier Encryption.

36 Ibid., sec. 3.8.1 XMLEncryption-based NameIdentifier Encryption, line 2092, Ibid., sec. 3.8.1 XMLEncryption-based NameIdentifier Encryption, line 2066.

37 Ibid., sec. 3.6. Identity Provider Introduction.

38 Ibid., sec. 3.6.1. Common Domain Cookie.

39 Cantor, Scott (Ed.), et al., *Liberty ID-FF Bindings and Profiles Specification,* http://www.projectliberty.org/specs/draft-liberty-idff-bindings-profiles-1.2-errata-v2.0.pdf., Ibid., sec. 2.1. SOAP Binding for Liberty.

40 Ibid., sec. 2.1.2.1. Authentication, Ibid., sec. 2.1.2.2. Message Integrity, 2.1.2.3. Message Confidentiality, Ibid., sec. 2.1.2.4 Security Considerations.

41 Aarts, Robert (ed.), et al., *Liberty ID-WSF Authentication Service and Single Sign-On Service Specification,* http://www.projectliberty.org/specs/draft-liberty-idwsf-authn-svc-v2.0-02.pdf.

42 Ibid., sec. 4. Authentication Protocol.

43 Myers, John G., *Simple Authentication and Security Layer (SASL), http://www.ietf.org/rfc/rfc2222.txt?number=2222, 1997.*

44 Aarts, Robert (Ed.), et al., *Liberty ID-WSF Authentication Service and Single Sign-On Service Specification,* http://www.projectliberty.org/specs/draft-liberty-idwsf-authn-svc-v2.0-02.pdf, sec. 5. Authentication Service.

45 Aarts, Robert (Ed.), et al., *Liberty ID-WSF Authentication Service and Single Sign-On Service Specification,* http://www.projectliberty.org/specs/draft-liberty-idwsf-authn-svc-v2.0-02.pdf, sec. 6. Single Sign-On Service.

46 Beatty, John (Ed.), et al., *Liberty ID-WSF Discovery Service Specification,* http://www.projectliberty.org/specs/draft-liberty-idwsf-disco-svc-v2.0-02.pdf.

47 Aarts, Robert (Ed.), et al., *Liberty ID-WSF Interaction Service Specification,* http://www.projectliberty.org/specs/draft-liberty-idwsf-interaction-svc-v2.0-01.pdf.

48 Ibid., Figures 1,2, and 3.

49 Ibid., sec. 3. The UserInteraction Element.

50 Ibid., sec. 4. The RedirectRequest Protocol.

51 Ibid., sec. 4.1. The RedirectRequest Element.

52 Ibid., sec. 5. Interaction Service.

53 Ibid., sec. 5.1.1. The InteractionRequest Element, Ibid., sec. 5.2.1. The InteractionResponse Element.

54 Ellison, Gary (Ed.), et al., *Liberty ID-WSF Security Mechanisms,* http://www.projectliberty.org/specs/draft-liberty-idwsf-security-mechanisms-core-v2.0-03.pdf.

55 Linn, John (Ed.), et al., *LibertyTrust Models Guidelines,* http://www.projectliberty.org/specs/liberty-trust-models-guidelines-v1.0.pdf., Varney, Christine (Ed.), et al., Project Liberty Privacy and Security Best Practices, http://www.projectliberty.org/specs/final_privacy_security_best_practices.pdf.

56 Landau, Susan (Ed.), et al., *Liberty ID-WSF Security and Privacy Overview,* http://www.projectliberty.org/specs/liberty-idwsf-security-privacy-overview-v1.0.pdf.

57 Ellison, Gary (Ed.), et al., *Liberty ID-WSF Security Mechanisms,* http://www.projectliberty.org/specs/draft-liberty-idwsf-security-mechanisms-core-v2.0-03.pdf, sec. 5.1. Transport Layer Channel Protection.

58 Ibid., sec. 5.2. Message Confidentiality Protection.

59 Ibid., sec. 6.1. Authentication Mechanism Overview.

60 Ibid., sec. 6.2.1. Unilateral Peer Entity Authentication, Ibid., sec. 6.2.2. Mutual Peer Entity Authentication.

61 Ibid., sec. 6.3. Message Authentication, Ibid., sec. 6.3.1. X.509mv3 Certificate Assertion Message Authentication, Ibid., sec. 6.3.2. SAML Assertion Message Authentication, Ibid., sec. 6.3.3. Bearer Token Authentication.

62 Aarts, Robert (Ed.), *Liberty ID-WSF Profiles for Liberty enabled User Agents and Devices,* http://www.projectliberty.org/specs/liberty-idwsf-client-profiles-v1.0.pdf.

63 Aarts, Robert (Ed.), *Liberty Reverse HTTP Binding for SOAP Specification,* http://www.projectliberty.org/specs/draft-liberty-paos-v2.0-01.pdf.

64 Ibid., sec. 6. Indication of Binding Support.

65 Ibid., sec. 7. Supported Message Exchange Patterns.

66 Hodges, Jeff (Ed.), et al., *Liberty ID-WSF SOAP Binding Specification,* http://www.projectliberty.org/specs/draft-liberty-idwsf-soap-binding-v2.0-01.pdf.

67 Ibid., sec. 5.1. Message Correlation: The <Correlation> Header Block.

68 Kainulainen, Jukka (Ed.), et al., *Liberty ID-WSF Data Services Template Specification,* http://www.projectliberty.org/specs/draft-liberty-idwsf-dst-v2.0-06.pdf., Ibid., sec. 4. Querying Data, Ibid., sec. 5. Modifying Data.

69 Kainulainen, Jukka (Ed.), et al., *Liberty ID-WSF Data Services Template Specification,* http://www.projectliberty.org/specs/draft-liberty-idwsf-dst-v2.0-06.pdf, sec. 6. Subscriptions and Notifications.

70 Kellomäki, Sampo (Ed.), et al., *Liberty ID-SIS Personal Profile Service Specification,* http://www.projectliberty.org/specs/liberty-idsis-pp-v1.1.pdf.

71 Kellomäki, Sampo (Ed.), et al., *Liberty ID-SIS Employee Profile Service Specification,* http://www.projectliberty.org/specs/liberty-idsis-ep-v1.1.pdf.

Chapter 7: Enabling Mobile Web Services

Bibliography

[Series 80 SDK NWSF API Reference] http://forum.nokia.com

[Series 80 NWSF developer Guide] http://forum.nokia.com

[Nokia WSDL Converter] http://forum.nokia.com

[Liberty ID-WSF Specifications] http://www.projectliberty.org/specs

[Nokia WS Web page] http://www.nokia.com/webservices

[Symbian OS Explained] Symbian OS Explained – Jo Stichbury, Wiley (an excellent description of Symbian client-server and active objects)

References

1 *It is also possible to discover a service (e.g., via a UDDI service registry or by pre-provisioning the service consumer application) before being authenticated for it.*

2 *This example deliberately omits the typical Symbian memory management code (CleanupStack::PushL() and Pop()) for clarity.*

Chapter 9: Java Client Development

1 *Developer's Guide to Nokia Web Services Framework for Java. DN0546933. The guide is located in the documentation directory of an installed Nokia Enhancements for Java™ for Series 80 SDK. The directory also contains the JavaDocs (Nokia Web Services Framework for Java API JavaDocs).*

2 *Liberty Alliance Project Web site <http://www.projectliberty.org>.*

3 *JSR-36: J2ME™ Connected Device Configuration*

Java Community Process (http://www.jcp.org/en/jsr/detail?id=36)

4 *JSR-46: J2ME™ Foundation Profile*

Java Community Process (http://www.jcp.org/en/jsr/detail?id=46)

5 *JSR-62: J2ME™ Personal Profile Specification*

Java Community Process (http://www.jcp.org/en/jsr/detail?id=62)

6 *Nokia Enhancements for Java for Series 80*

Forum Nokia (http://www.forum.nokia.com)

7 *Series 80 Developer Platform 2.0 SDK for Symbian OS – For Personal Profile*

Forum Nokia (http://www.forum.nokia.com.)

Chapter 10: C++ Client Development

Resources

Programmer's Guide To The Service Development API

\Nokia\Tools\Web_Services_Enhancement_for_Series_80\Docs

Service Development API

\Nokia\Tools\Web_Services_Enhancement_for_Series_80\Docs\Nokia_Service_Development_API.chm

Symbian OS: Getting Started with C++ Application Development

\Symbian\7.0s\S80_DP2_0_PP_SDK\Series80Doc

Nokia WSDL-to-C++ Installation Guide

References

[LIBERTY] http://www.projectliberty.org

[WS-ARCH] http://www.w3.org/TR/2004/NOTE-ws-arch-20040211/

[NOKIA-WS] http://www.nokia.com/webservices

[WSDL] http://www.w3.org/TR/wsdl

[SASL] http://www.ietf.org/rfc/rfc2222.txt

[SOAP] http://www.w3.org/TR/soap12/

[ID-WSF] http://www.projectliberty.org/specs/index.html

[WSDL2C] Nokia WSDL-to-C++ Wizard, User's Guide for Visual Studio

[WSS] http://www.oasis-open.org/specs/index.php#wssv1.0 [WS-I] http://www.ws-i.org/

Appendix A: Web Services Standards Organizations

[OASIS] http://www.oasis-open.org/ , Accessed 3 January 2005.

OASIS. About OASIS. <http://www.oasis-open.org/who/>. Accessed 3 January 2005.

[Cover-Pages] http://xml.coverpages.org/ , Accessed 3 January 2005.

[XML.org] http://www.xml.org/, Accessed 3 January 2005.

[ebXML.org] http://www.ebxml.org/, Accessed 3 January 2005.

WS-I. Interoperability: Ensuring the Success of Web Services. WS-I Overview Presentation. <http://www.ws-i.org/docs/20041130.introduction.ppt>. Accessed 3 January 2005.

WS-I. How we work. <http://www.ws-i.org/about/workhow.aspx>. Accessed 3 January 2005.

[ProjectLiberty] http://www.projectliberty.org

W3C. About the World Wide Web Consortium (W3C). <http://www.w3.org/Consortium/>.

Accessed 17 January 2005. For permission to quote and paraphrase, see W3C Document Notice and License included in this Appendix, and available at <http://www.w3c.org/Consortium/Legal/copyright-documents-19990405>.

W3C. Device Independence. <http://www.w3.org/2001/di/>. Accessed 17 January 2005.

W3C. Technical Architecture Group. <http://www.w3.org/2001/tag/>. Accessed 4 January 2005.

W3C. Technical Reports and Publications. <http://www.w3.org/TR/>. Accessed 5 January 2005.

W3C. Open Source Software. <http://www.w3.org/Status.html>. Accessed 5 January 2005.

W3C. [W3C-Patent-Transition] http://www.w3.org/2004/02/05-pp-transition, Accessed 6 January 2005.

W3C. Patent Policy. <http://www.w3.org/Consortium/Patent-Policy/>. Accessed 6 January 2005.

Appendix B: Nokia Web Services Development API – Quick Reference

1 *The APIs depicted here are the native Symbian C++ libraries. The Java interfaces map to these libraries directly but may differ in minute differences, specifically due to differences in the programming language environment. The behavior, though, of equivalent methods can be assumed the same.*

Index

C

circle of trust 106

E

Extensible Markup Language. *See XML*

F

Forum Nokia 124
Forum Nokia Pro 124

I

ID-FF 104
  architecture 108
  Profiles 112
  Protocols 109
  Single Logout 110
  Single Sign-On and Federation 109
ID-FF Bindings
  definition 114
ID-FF Profiles
  description 112
  Identity Federation Termination Notification 113
  Identity Provider Introduction 114
  NameIdentifier Encryption 114
  NameIdentifier Mapping 113
  Register Name Identifier Profiles 113
  Single Logout 113
  Single Sign-On and Federation 112
ID-FF Protocols
  description 109
  Federation Termination Notification 110
  Name Identifier Mapping 111
  Name Registration Protocol 110
  Single Logout 110
  Single Sign-On and Federation 109
ID-SIS 104,122
  description 122
  Employee Profile Service 122
ID-WSF 104
  Client Profiles 120
  practical example 115
  SOAP Binding 121
ID-WSF Service Specifications
  Authentication Service 117
  Authentication Service and Single Sign-On Service Specification 116
  Data Services Template 121
  Discovery Service 117
  Interaction Service 117,118
  Security Mechanisms 119
  Single Sign-On Service 117
Identity
  achieving direct authentication 83
  and mobile Web services 82
  and Web Services 81
  Audit 89
  authentication assertion 84
  authentication context 85
  Authentication of a mobile device 84
  Authorization 89
  circle of trust 106
  client authentication 84
  determining used authentication technique 84
  discovery service 91
  Employee Profile Service 122
  End-to-End Web service client authentication 83
  establishing 82
  factors 82
  federation 103,105
  HTTP basic authentication 83
  ID-FF 104
  ID-SIS 104
  ID-WSF 104
  identity federation 88
  legal statute 82
  Liberty Alliance Identity Technologies 103
  Liberty Discovery Service 91
  peer authentication 83
  Personal Profile Service 122
  practical example 105
  replay attacks 86
  role of IDP 84

SAML 2.0 85
SAML token 84
services using identity information 90
service description 91
Single Sign-On 87
Single Sign-On Enabled Web Services 88
SOAP Message Security 83
SSL/TLS 83
standards portfolio 103
Web service authentication tokens 85
X.509 certificate 83
Identity Web Services
Connecting to 138
discovery 138
privacy considerations 138
SOA for S60 platform 138
steps taken when connecting 139

L

Liberty Alliance
Achievements 276
Deliverables 275
Future Directions 278
Influence 278
Membership Base 271
Operating Model 272
Organization 272
role 271
Value Proposition 271
Liberty Alliance Identity Technologies
circle of trust 106
ID-FF 104
ID-SIS 104
ID-WSF 104
Liberty Identity Federation Framework (ID-FF) 104
Liberty Identity Services Interface Specifications (ID-SIS) 104
Liberty Identity Web Services Framework (ID-WSF) 104
Liberty Identity Web Service Framework (ID-WSF) 115
major elements 104
practical example 105
rationale for 103
role in the Liberty architecture 115
Liberty Identity Federation Framework. *See* ID-FF
Liberty Identity Services Interface Specifications. *See* ID-SIS
Liberty Identity Service Interface Specifications (ID-SIS) 122
Liberty Identity Web Services Framework. *See* ID-WSF

M

markup languages
usage 7
Mobile devices
limitations of 125
operating systems 126
mobile Web services
architecture 3
challenges 2
characteristics 101
major application areas 2
requirements 2
Software Development Environment 125

N

Nokia Web Services Framework. *See* NWSF
NWSF
Active Objects 127
API 166
availability 123
Client-Server Architecture 126
for Series 80 219
HTTP Stack 127
Java architecture 127
Java code to access the HelloWS Web service 134
Java support 123
XML Processing 127,168
NWSF C++ Client Development
adding IDPs to NWSF 240
and WSDL 224
Associating an Identity Provider with a Service 240
associating IDPs with services 239
Callbacks 248
Configuration 244
connecting to a Web service without a framework 238
Creating an XML handler 232
Creating a Query 250
deactivating frameworks 238

Description of wscexample Client Code 246
DOM interface 223
errors 235
Establishing a Service Connection 247
frameworks 237
framework identifier 237
full SOAP envelopes 232
HandleError() 235
HandleMessage() 231
Handling Callbacks 248
HelloWS.wsdl document 225
HelloWS main client code 228
HelloWS messages 227
Hello Web Service Example 226
HTTP Basic Authentication 239,242
Identity 239
IDE Usage 223
IDP-Based Authentication 241
IDP identifier 240
Linking a service provider with an IDP. 240
MSenServiceConsumer Interface 230
Nokia Service Development API 222
parsing an incoming XML message 224
PhonebookEx Example Application 252
Policy 243
Registering an Identity Provider 240
registering credentials for HTTP Basic Authentication 247
Registering Service Policy 244
returning a full SOAP envelope 232
sample applications 219
SAX parsing 223
selecting individual pieces of XML information 224
self-authentication 239
sending a message 230
sending a query with a synchronous method 250
SenIdentities.xml 245
SENSERVCONN.lib 222
SENSERVDESC.lib 222
SenServDesc library 245
SENSERVMGR.lib 222
SenSessions.xml 245
SENUTILS.lib 222
SENXML.lib 222
SENXMLINTERFACE.lib 222
Service Connection 222
service contract 240
Service Database 239
Service Description 222
Service descriptions 239
Service Description Library 238
Service Development API 222
Service Development API general package structure 222
Service Manager 222,239
Service Manager Library 238
SetStatus() 236
submitting messages synchronously 237
Test Service Code 251
Username Token 239,241
using a framework to access a Web service 237
Using HTTP Basic Authentication 242
Using IDP-Based Authentication 241
Using WS-Security Username Token 241
using WSDL 224
Utilities 222
Utility Library 245
WS-Security 239,241
wscexample application 246
WSDL-to-C++ transformation 226
XML Library 223
XML Processing 222

NWSF for Java
NWSF API 166
Overview of architecture 166

NWSF Java Client Development
and identity 178
and identity providers 178
and IDP-based authentication 179
and security 178
and Self-Authentication 186
associating Identity Providers with another services 178
classes in the XML package 168
com.nokia.mws.wsf.IdentityProvider 166
com.nokia.mws.wsf.ServiceConnection 166
com.nokia.mws.wsf.ServiceDescription 166
com.nokia.mws.wsf.ServiceManager 166

com.nokia.mws.wsf.WSFObjectFactory 167
com.nokia.mws.wsf.xml 168
com.nokia.mws.wsf.xml.Element 167
Configuration of Service Development API 167
Configuring framework plug-ins 176
Connecting to Web Services 175
connecting to Web services by using a framework 175
Connecting to Web Services without Using a Framework 176
Creating New XML Messages 169
Development Tools 165
Frameworks 176
HelloWS.java 189
HelloWS.java Test Service Servlets 193
HelloWS Example 172
HelloWS example message exchange 173
HTTP Basic Authentication 186
interfaces in the XML package 168
linking an IDP with an SP 179
NWSF API 166
Parsing Incoming XML Messages 170
Policy 189
Prerequisites 164
Registering an Identity Provider 178
registering a service 177
SenSessions.xml 167
service database 177
Username Token 185
using a basic Web service 189
using a Liberty-compliant Web service 189
using HTTP Basic Authentication 189
using IDP-based authentication 179
Using the Service Database 177
WS-Security 185
XML handler implementation characteristics 169
XML Processing 168

O

OASIS
Achievements 264
Deliverables 264
Future Directions 266
Influence 266
Membership Base 263
Operating Model 263
Organization 263
role 262
Value Proposition 263

P

PAOS
description 92
policy
and SOA for S60 platform 148
authorization 72
definition 71
domain-specific 73
interoperability 74
types 72

R

Reverse HTTP Binding for SOAP 120

S

Security
and Web services 94
Audit 94
Authentication 94
Authorization 94
Confidentiality 94
cryptography 96
Denial-of-service attack mitigation 94
encryption 100
HMAC 97
Integrity 95
Integrity protection 99
mechanisms 95
message confidentiality 100
message digest 97
message integrity 97,99
mitigating Denial-of-Service attacks 101
mitigating the risks 94
protocol layer interactions 95
Replay protection 95
SAML token 148
signature algorithms 96
single-hop communication 96
SSL/TLS 96
Training 95
WS-Security 99
XML Encryption 95,100

XML Signature 95
Service-Oriented Architecture
benefits 26
Description 22
description 21
Discovery 22
interoperability 26
major components 22,23
Messaging 22
service consumer 21
service provider 21
with Web Services 23
service description
explanation 22
WSDL 22
service discovery
description 22
service messaging
description 22
SOAP 22
Simple Object Access Protocol. *See* SOAP
Single Sign-On
and Web Services 88
description 87
SOAP
actors 28
and XML Schema 29
application payload 28
attachments 32
binary content 27
body 27
description 27
document messaging model 29
envelope 27
envelope structure 27,28
example of an intermediary 31
example request and response 30
header 27
header blocks 28
header usage 28
initial SOAP sender 31
intermediaries 31
Message Security 83
Multipart-MIME 32
roles 28
RPC model 29
security header 28
SOAP over HTTP 34
transport 33
ultimate SOAP receiver 31
validating element types 29
validating the content structure 29
XML namespaces 27
SOA for S60 platform
Active Objects 127
adding frameworks 130
and identity providers 136
and self-authentication 135
architecture 128
associating a service with a framework 132
asynchronous access to a Web service 133
authentication 135
Authentication Service client 130
availability 123
Choosing an XML Parsing Strategy 158
Client-Server Architecture 126
common framework for authentication 130
Connecting to an Identity Web Service 138
connecting to an identity Web service 139
Discovery Service client 130
establishing a service connection 131
Example message flow using an identity service 143
Frameworks 129
framework plug-in 130,149
framework plug-in availability 130
HelloWS example 131
HTTP Basic Authentication 135
HTTP Basic Auth example message flow 136
HTTP Stack 127
Identity Providers 146
Identity Web Services Support 138
Java architecture 127
Java support 123
Liberty ID-WSF framework plug-in 129
network endpoint 132
OASIS Web Services Username Token 135
pattern facets 147
Pluggable Architecture 149
pluggable interface 130

Policies 148
policy statements 132
Quick Reference 283,313
responsibilities of frameworks 130
role of service database in connection creation 144
schema contract 132
Service Association 145
Service Connection Library 284
Service Database 144
Service Database Service Association 145
service description 132
Service Description Library 286
Service Manager 126,129,130
Service Manager Library 285
Service Matching 147
Service Policies 148
Service Providers 148
SOAP/HTTP message flow example 133
synchronous access to a Web service 133
Utilities Library 294
Web Service Consumer 129
Web Service Provider 129
Web Service Providers 148
Writing Code for 158
XML API 152
XML Library 290
XML Processing 127,150
XML processing features 153

Standards Organizations
Liberty Alliance 271
OASIS 262
W3C 257
WS-I 267

U

UDDI
bindingTemplate element 47
businessEntity element 47
businessService element 47
classification 45
data models 47
description 45
general characteristics 45
identifiers 45
Inquiry API 48
inquiry patterns 48
Programmers APIs 48
Publication API 48
publisherAssertion element 47
Registry 46
Registry Categorization 46
Subscription API 48
subscription element 48
technical service information 46
tModel element 47

Universal Description, Discovery, and Integration. *See* UDDI

W

Web Services Description Language. *See* WSDL

WSDL
Binding 41
Description 35
Document Structure 38
Message 39
Port 43
PortType 40
Technical Details 36
Types 38

X

XML 232
and Web services 7
attributes 9
attribute namespace qualification 12
capabilities of DTD 13
Choosing a Parsing Strategy 158
CSenBaseElement 153
CSenBaseFragment 156
CSenDomFragment 154
declaration 9
default namespaces 11,12
defining instance structure and syntax 14
description 7
Document Type Definition (DTD) 13
DOM 153
DOM interface 223
DTD rules 13
elements 8
hierarchical data structures 8
importing XML Schema files 19
including XML Schema files 19
incorporating external XML Schema data 17

markup 8
namespaces 9
namespace definition 11
namespace inheritance 11
namespace prefix 11
namespace usage scenarios 10
node tree 153
parsing XML 150,152
practical usage 8
qualifying locally declared elements and attributes 20
rationale for namespaces 10
SAX parsing 223
SOA for S60 Platform 150
specification 8
specifying document structure 13
specifying legal data types 13
target namespace 18
valid XML documents 13
well-formed XML documents 13
xmlns attribute 11
XML namespaces 27
XML Schema 14
XML Schema complex types 15,16
XML Schema model groups 15
XML Schema namespaces 17
XML Schema simple types 14
XML Schema types 14
XML Schema
combining multiple documents 17
complex types 15
datatypes 14
features 14
importing XML Schema data 19
including XML Schema data 19
model groups 15
namespaces 18
number of element occurrences 16
occurrence of attributes 16
qualifying locally declared attributes 20
qualifying locally declared elements 20
Simple Types 14
target namespace 18
using a simple schema type 15